POLYHEDRON MODELS FOR THE CLASSROOM

POLYHEDRON MODELS

for the Classroom

Second Edition

MAGNUS J. WENNINGER

St. Augustine's College
Nassau, Bahamas

NCTM
NATIONAL COUNCIL OF
TEACHERS OF MATHEMATICS
1906 Association Drive, Reston, VA 20191-1593

Eighth printing 1999

ISBN 0-87353-083-7

Printed in the United States of America

Contents

Photographs: Stanley Toogood's International Film Productions, Nassau, Bahamas

Introduction

This booklet, first published in 1966, went through its sixth printing in 1973, making up a total of 35 000 printed copies. That fact alone attests to the continuing interest in this work on the part of teachers and students alike.

The second edition now being presented adds a new feature—seven center-fold sheets containing full-scale drawings from which nets or templates may be derived to make the models shown in the photographs and described in the text.

The following procedure is recommended for making the templates. Carefully loosen the staples at the center of the booklet and remove the center-fold pages. Having done this, turn the staples back down so that the booklet remains intact for reading and reference. The center-fold pages may now be cut with scissors. Leave a quarter-inch margin all around the edge of each pattern. This will provide the proper net or template for making the models.

If you do not wish to cut the center-fold pages, or if you would like to make a more durable net or template, you may copy the pattern onto heavier paper. First, place the pattern over a sheet of heavier paper (perhaps index card stock) and then prick through the vertices of the pattern with a divider point or a needle. Use a straightedge to draw the outline of the pattern onto the heavy paper and trim as described before, leaving a quarter-inch margin all around the edge. This net or template is to be used to multiply parts, which are treated as described on page 5. *Do not trim the net or template.* Trim only the parts to be used for making the model. Specific directions are given for trimming the parts shown on pages 4, 13, 15, and 26. It is assumed that your own experience and ingenuity will dictate what is best in all other cases. The templates may then be saved and used for making more models of the same kind or, in some instances, for making different models where a particular template is used in combination with others.

If you should need or want more explicit instructions for making the models described on pages 28–40, you will find them in a larger work of mine, a book entitled *Polyhedron Models,* published by Cambridge University Press (New York, 1974) in paperback form. This is a source book for all 75 known uniform polyhedra and for some stellated forms.

It might be of interest for readers to know that a definitive enumeration of uniform polyhedra has been made. John Skilling, of the Department of Applied Mathematics and Theoretical Physics at Cambridge University, has just published the results of his computer analysis of the problem. He shows that "the list of Coxeter *et al.* is indeed complete as regards uniform polyhedra in which only two faces meet at any edge. The natural generalization that any *even* number of faces may meet at an edge allows just one extra polyhedron to be included in the set." (See John Skilling, "The Complete Set of Uniform Polyhedra," *Philosophical Transactions of the Royal Society of London,* ser. A, vol. 278, no. 1278 [March 1975].)

M. J. W.

Nassau, Bahamas
April 1975

Introduction to the First Edition

The study of polyhedra is an ancient one, going back to the dawn of history. It is especially those polyhedra that are called uniform that have evoked the greatest interest and provided the most fascination. It should therefore be of special usefulness for mathematics students and teachers in their classrooms today to see and handle these geometrical solids in aesthetically pleasing models and to be delighted with their beauty and form.

Most students show immediate interest in this kind of work, and teachers are often surprised to see the quality of the results a student obtains in making the models. It is a genuine outlet for the creative instinct; in addition, it calls for care and accuracy, as well as perseverance and pertinacity in models that have many parts and an intricate color arrangement. It is also surprising how the models can stimulate interest in some of the basic theorems of solid geometry. And, when a project is finished, the models will enhance the appearance of the classroom, where they can be put on permanent display.

HEXAHEDRON, OR CUBE

OCTAHEDRON

THE FIVE PLATONIC SOLIDS

TETRAHEDRON

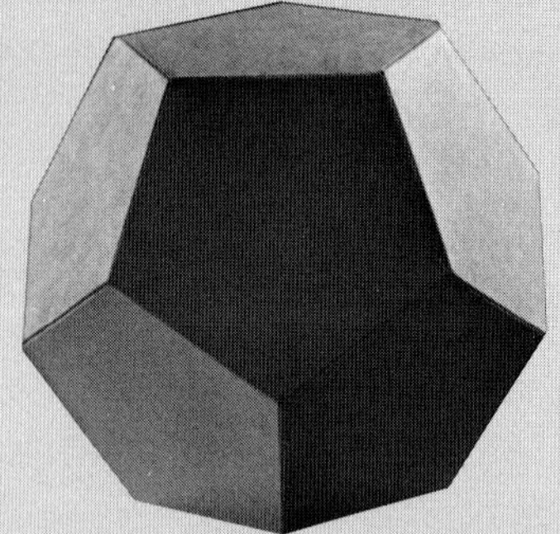
DODECAHEDRON

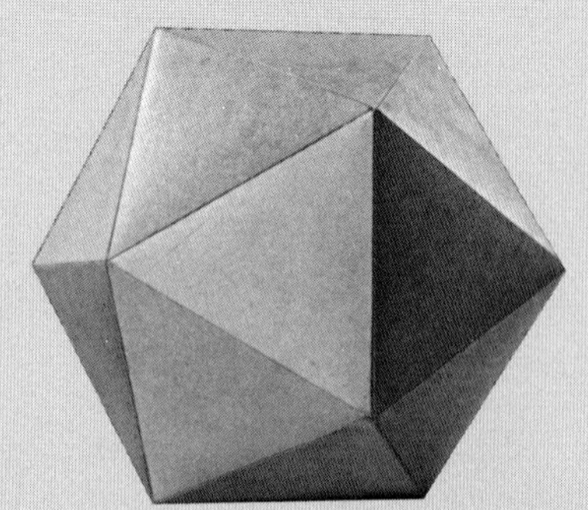
ICOSAHEDRON

The Five Platonic Solids

The most ancient polyhedra are the set of five known as the Platonic solids. They derive their name from the great Greek philosopher Plato, who discovered them independently about 400 B.C., though they were probably known before Plato. The ancient Egyptians knew four of them: the tetrahedron, octahedron, and cube are found in their architectural design, and Egyptian icosahedral dice are to be found in an exhibit in the British Museum. According to Heath, the Etruscans were acquainted with the dodecahedron before 500 B.C.[1] All five were studied by the early Pythagoreans before the time of Plato and Euclid. It is in the *Elements* of Euclid, however, that we find the most extensive treatment of the geometry of these five solids.

Today models of these solids, usually in plastic, are featured in the catalogues of scientific and educational supply houses. But models in heavy paper are so easily made and so useful as a project for students that it is well worth the effort to make a set. The nets, or patterns, for making these models are given in many geometry textbooks. It will be found that the models are even more attractive when they are made with facial polygons of various colors. (Suggested color patterns are set out in Figures 1, 2, 3, 4, and 5.)

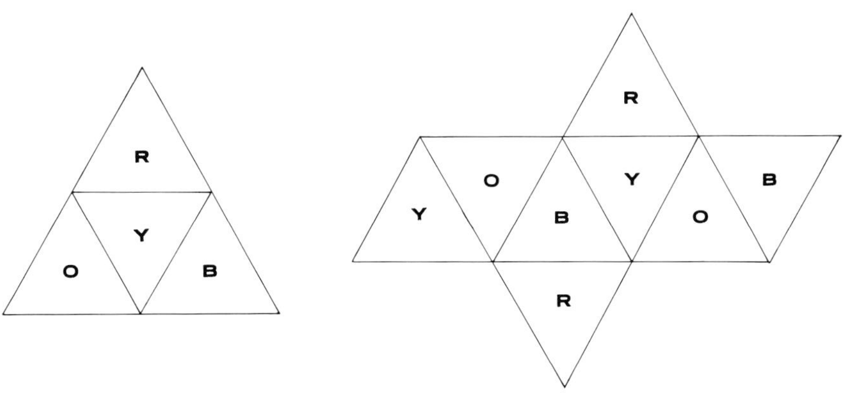

FIG. 1 FIG. 2

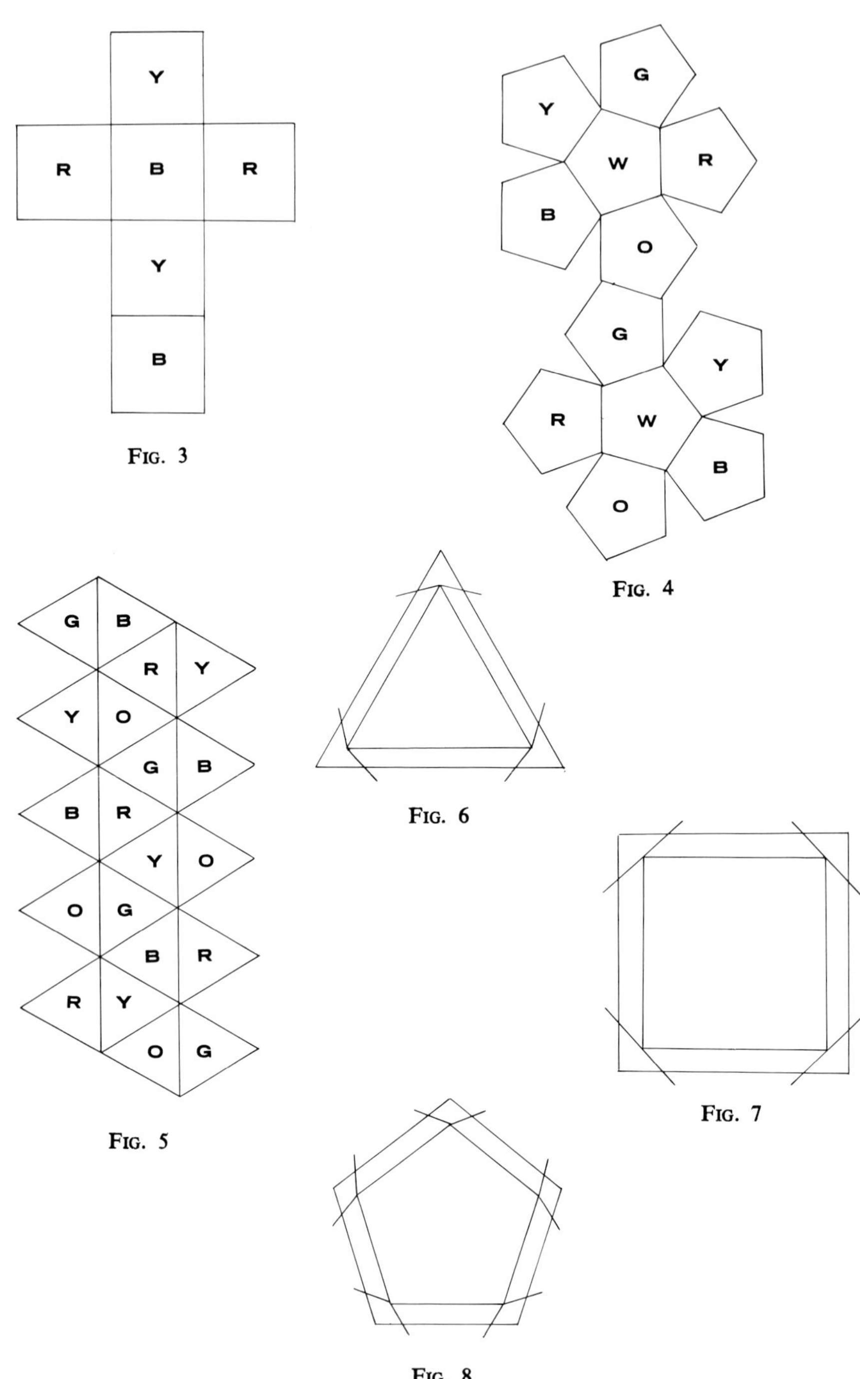

FIG. 3

FIG. 4

FIG. 5

FIG. 6

FIG. 7

FIG. 8

For easiest construction and sharpest edges, the material to use is heavy paper with a somewhat hard finish—the type used for file cards. It can be bought in larger sheets, in colors, under the name "colored tag." Pastels are very suitable, and they are a good alternative to deeper colors. In the method of construction suggested here you need only one triangle, square, or pentagon as a net. Three, four, or five sheets of colored paper may be stapled together and the net placed on top of them. Then, using a sharp needle and pricking through all the sheets at one time, make a hole in the paper at each vertex of the net, which is held as a guide or template. In this way exact copies of each part are quickly obtained. Next give the paper an initial trimming with scissors, with all the sheets still held firmly together by the staples. You must be careful to provide about a quarter-inch margin all around to be used for flaps or tabs to cement the parts together. After this it will be best to treat each part individually. Experience will soon teach you that the accuracy of your completed model is directly proportional to the care you have lavished on each individual part. With a sharp point, such as that of a geometry compass, you must now score the paper, using a straightedge or set square as a guide to connect the needle holes with lines. (Pencil lines are not needed, since the process of scoring sufficiently outlines the shape of each part.) More accurate trimming is next to be done. (For suggestions on how to do this, see Figures 6, 7, and 8.) The scored lines then make folding of the tabs a simple and accurate operation.

A good household cement provides the best adhesive, since it is strong and quick-drying. A few wooden clothespins of the coiled-spring variety that have been turned inside out make excellent clamps. When these are used, the cementing can proceed rapidly and the clamps can be moved from one part to the next in almost a matter of minutes as each part is successively cut and trimmed. The last part will not give you too much trouble if you first cement one edge and let it set firmly, then proceed to put cement on the other edges and close down the last polygon as you would close the lid on a box. A needle or compass point makes a good instrument to maneuver the last edges into accurate position. Deft fingers and a little practice will do the rest.

The Thirteen Archimedean Solids

Once a set of the five Platonic solids has been made, the next project will certainly be to make a set of the thirteen Archimedean solids. These too have an ancient history. Plato is said to have known at least one of them, the cuboctahedron. Archimedes wrote about the entire set, though his book on them is lost. Kepler is the first of the moderns to have treated these solids in a systematic way. He was also the first to observe that two infinite sets of polyhedra, the set of prisms and the set of antiprisms, have something in common with the thirteen Archimedeans, namely, membership in the set known as the semiregular polyhedra.[2] (A semiregular polyhedron is one that admits a variety of polygons as faces, provided that they are all regular and that all the vertices are the same.)

As in the case of the Platonic solids, so too in that of the Archimedeans the beauty of the set is greatly enhanced by suitable color arrangements for the faces. Since it is evident that many different color arrangements are possible, you may find it interesting to work out a suitable arrangement for yourself. The general principle is to work for some kind of symmetry and to avoid having adjacent faces with the same color. This may remind you of the map-coloring problem. The fact is that a polyhedron surface is a map, and as such is studied in the branch of mathematics known as topology. In making these models, however, you need not enter into any deep mathematical analysis to get what you want. Your own good sense will suggest suitable procedures. (See page 8.)

The actual technique of construction is the same here as that given above: namely, only one polygon—a triangle, square, pentagon, hexagon, octagon, or decagon—will serve as a net. However, it is important to note that in any one model all the edges must be of the same length. If you want to make a set having all edges equal, you will find the volumes growing rather large with some models in the set. Of course a large model takes up more display space, so you must gauge your models with that fact in mind. On the other hand, you

TRUNCATED TETRAHEDRON

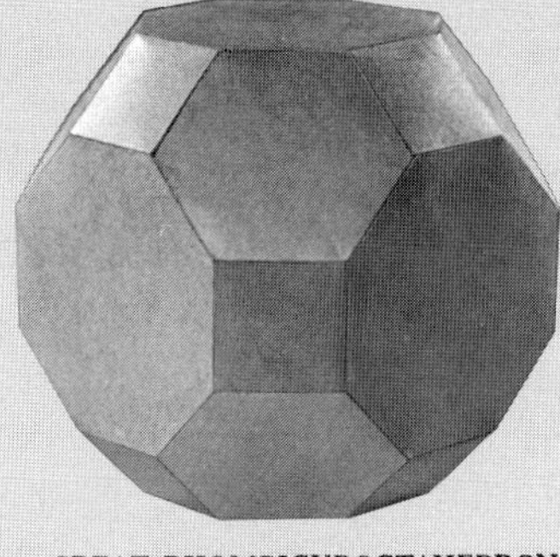

GREAT RHOMBICUBOCTAHEDRON

TRUNCATED DODECAHEDRON

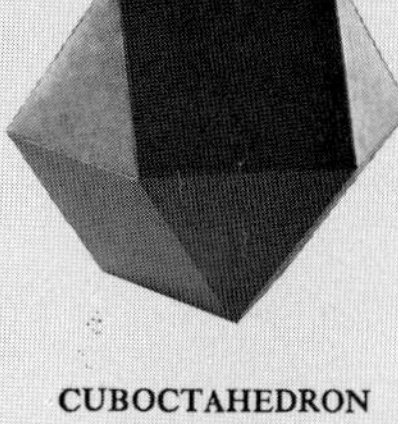

CUBOCTAHEDRON

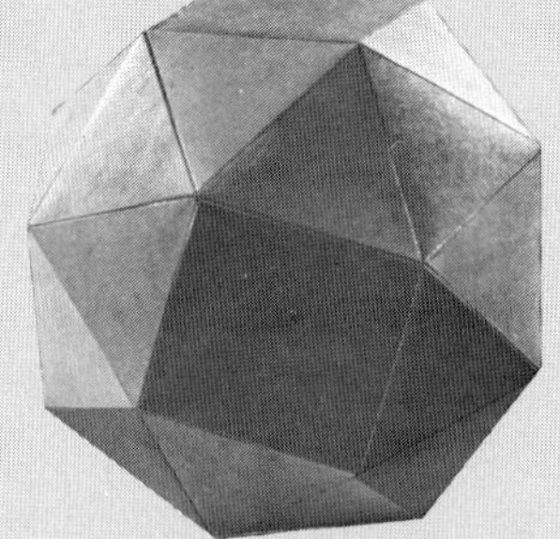

SNUB CUBE

TRUNCATED ICOSAHEDRON

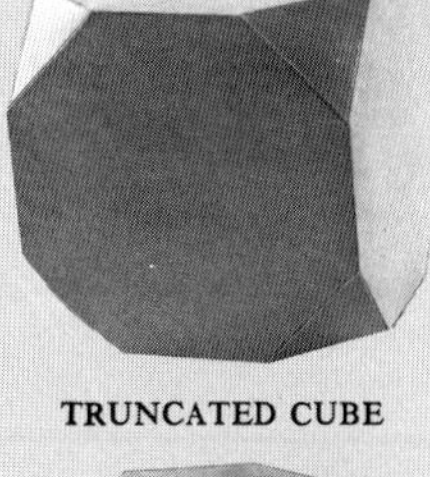

TRUNCATED CUBE

TRUNCATED OCTAHEDRON

SNUB DODECAHEDRON

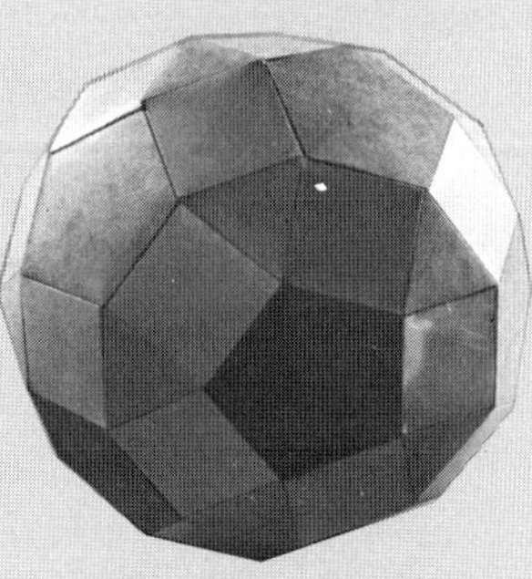

RHOMBICOSIDODECAHEDRON

RHOMBICUBOCTAHEDRON

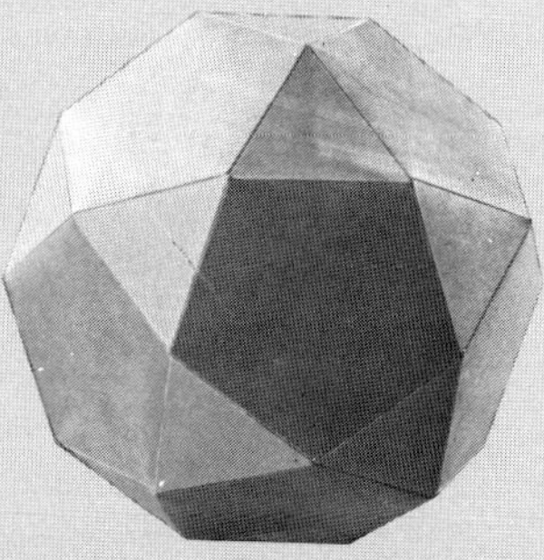

ICOSIDODECAHEDRON

GREAT RHOMBICOSIDODECAHEDRON

THE THIRTEEN ARCHIMEDEAN SOLIDS

may want to vary the edge length from model to model and thus obtain polyhedra of more or less uniform volume or actually of uniform height. Here experiment is in order, and a student can have an excellent demonstration appealing to his own experience of the geometrical theorems on the relation of similar figures or solids: linear dimensions are directly proportional to each other; areas are proportional to squares on linear dimensions; volumes, to cubes on linear dimensions.

Suggested Color Patterns for the Archimedean Solids

(R = rose, Y = yellow, B = blue, O = orange, G = green, W = white)

1. Truncated tetrahedron:
   - 4 hexagons G R B Y
   - 4 triangles G R B Y
2. Cuboctahedron:
   - 6 squares Y B R Y B R
   - 8 triangles G
3. Truncated cube:
   - 6 octagons G B R G B R
   - 8 triangles Y
4. Truncated octahedron:
   - 6 squares G
   - 8 hexagons R Y O B R Y O B
5. Rhombicuboctahedron:
   - 6 squares Y
   - 12 squares R
   - 8 triangles B
6. Great rhombicuboctahedron:
   - 6 octagons R Y B R Y B
   - 12 squares G
   - 8 hexagons O
7. Snub cube:
   - 6 squares Y R B Y R B
   - 8 triangles O
   - 8 triangles B
   - 8 triangles R
   - 8 triangles G
8. Snub dodecahedron:
   - 12 pentagons O
   - 15 triangles B
   - 15 triangles R
   - 15 triangles Y
   - 15 triangles G
   - 20 triangles O
9. Icosidodecahedron:
   - 3 pentagons R
   - 3 pentagons B
   - 3 pentagons O
   - 3 pentagons G
   - 20 triangles Y
10. Truncated dodecahedron:
    - 3 decagons Y
    - 3 decagons R
    - 3 decagons G
    - 3 decagons B
    - 20 triangles O
11. Truncated icosahedron:
    - 8 hexagons R
    - 6 hexagons G
    - 6 hexagons Y
    - 12 pentagons O
12. Rhombicosidodecahedron:
    - 12 pentagons R
    - 30 squares B
    - 20 triangles Y
13. Great rhombicosidodecahedron:
    - 12 decagons Y
    - 20 hexagons R
    - 30 squares B

Prisms, Antiprisms, and Other Polyhedra

The construction of a few prisms and antiprisms will next give you a good idea of why these belong to the set of semiregular solids. You already have the required nets, since they are the same polygons as those used for the Archimedean solids. A prism has any regular n-gon for end faces and squares for side faces. The antiprism has equilateral triangles instead of squares for side faces. Once you have made some of these, you will find that it is possible for you to branch off on your own into many different types of irregular solids. Some of these come up in textbook problems in a mathematics class: right pyramids (with regular bases or otherwise) having isosceles triangles for sides, various types of oblique pyramids, parallelepipeds, truncated pyramids, and others. You will have noticed that some of the Archimedeans are truncated versions of the Platonic solids. There are some interesting dissection problems in connection with the tetrahedron and the cube. These take on added interest for their usefulness as puzzles. If you are interested in mechanical drawing, you will find that many of the objects you are called upon to draw can also be constructed as models in paper according to the techniques suggested in this monograph. Sections of solids can also be illustrated in this way. In fact, even circular cones and conic sections can be similarly done. (See Figures 9-17 for some suggestions.) Then why not go ahead and discover some others on your own?

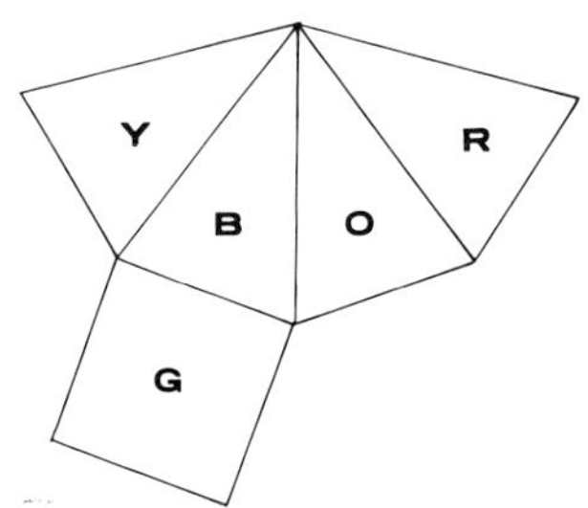

FIG. 9
Square pyramid

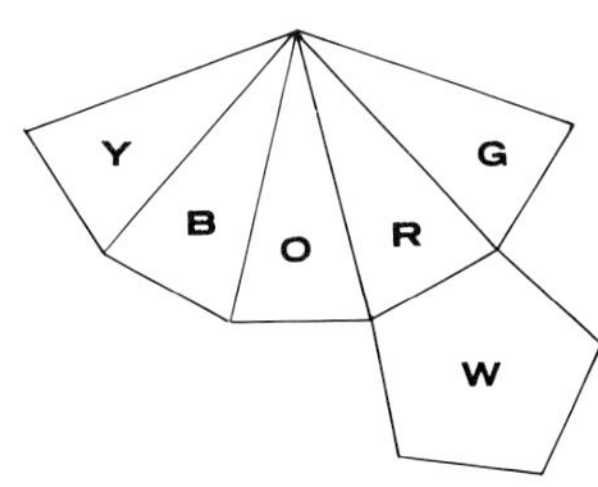

FIG. 10
Pentagonal pyramid

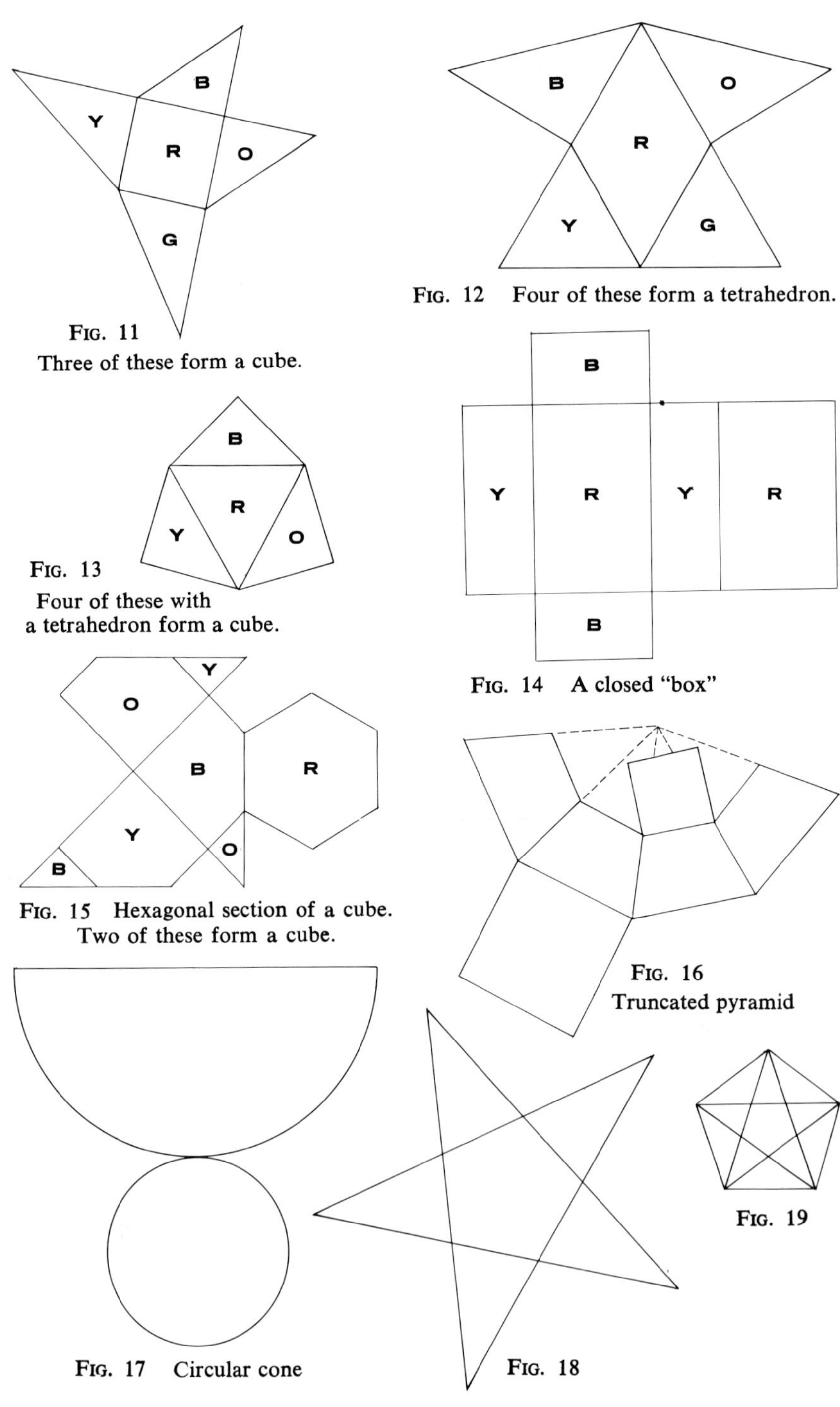

FIG. 11
Three of these form a cube.

FIG. 12 Four of these form a tetrahedron.

FIG. 13
Four of these with a tetrahedron form a cube.

FIG. 14 A closed "box"

FIG. 15 Hexagonal section of a cube. Two of these form a cube.

FIG. 16
Truncated pyramid

FIG. 17 Circular cone

FIG. 18

FIG. 19

The Four Kepler-Poinsot Solids

The next set of uniform polyhedra, four in number, whose beauty is most striking, is that of the Kepler-Poinsot solids. (Kepler discovered two about 1619, and Poinsot rediscovered these and discovered the two others in 1809.) These solids are all the more interesting because they were unknown to the ancient world. Of course the star polygon, also called the pentagram, is very ancient, possibly as old as the seventh century B.C.[3] The Pythagoreans used this five-pointed star as a symbol of their brotherhood. Yet the discovery of solids with star-shaped facial planes belongs to the modern era. It is perhaps correct to say that the ancients missed these because they were concerned only with convex polygons and polyhedra and did not consider the case of intersecting facial planes. These are the properties that enter into the solids to be described in what follows. It is precisely the star-shaped appearance of the Kepler-Poinsot solids that

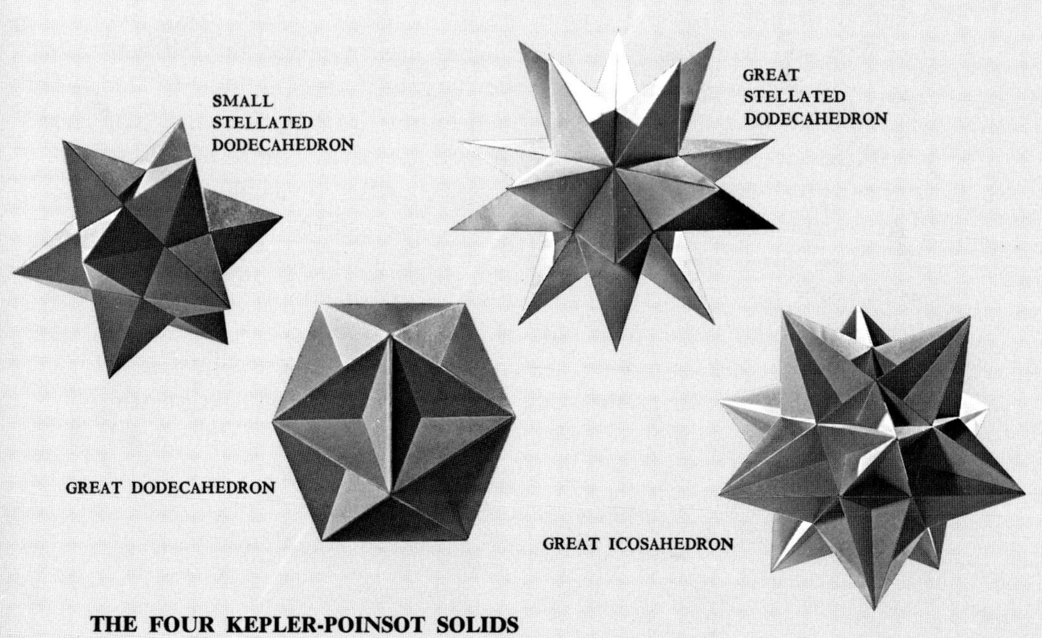

THE FOUR KEPLER-POINSOT SOLIDS

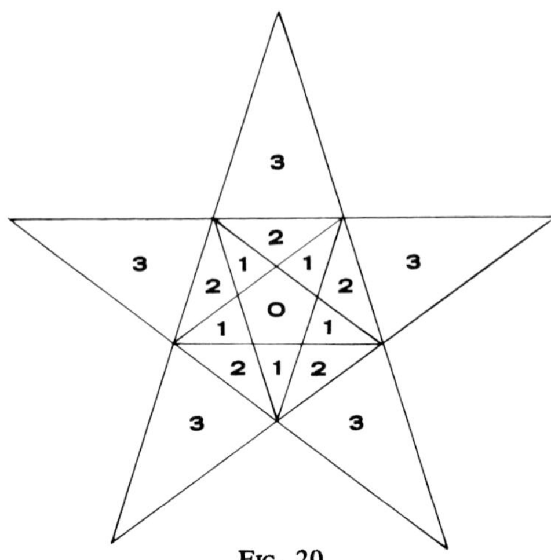

FIG. 20

gives them their interest and beauty—a beauty that, again, is greatly enhanced by making them so that each facial plane has its own color.

It was Cauchy who pointed out that these four polyhedra are actually stellated forms of the dodecahedron and icosahedron—the first three, of the former; and the last one, of the latter. This process of stellation is in itself a most interesting one. It is most readily understood by investigating the dodecahedron. First it may be noted that the five-pointed star, or pentagram, arises first by producing the sides of a pentagon or by drawing all the diagonals (see Figures 18 and 19). If both procedures are combined in one drawing, the result will give three sets of triangles, which provide the parts required for three of the Kepler-Poinsot solids: the small stellated dodecahedron, the great dodecahedron, and the great stellated dodecahedron (see Figure 20). The simplicity of this figure, providing such simple parts to be used as nets for the making of these solids, is something greatly to be cherished in these polyhedron models.

Before any description of the construction of these solids is set down, a further description of the stellation process will be helpful. If the facial planes of the dodecahedron, for example, are produced, those planes that intersect will generate certain cells: first a set of twelve pentagonal pyramids; then a set of thirty disphenoids, or wedge-shaped pieces; and finally a set of twenty triangular dipyramids.

Each of these sets of cells constitutes the external parts of successive stellations of the dodecahedron. In print this may sound complicated, but in models it is very evident.

As for constructing the models, you can do it without making the cells mentioned above. But if you are very ambitious, it would be an excellent project to make all the cells. You can easily discover the nets for these cells by yourself. Then you can show how the stellated forms arise by adding these cells to the basic solid in each case. In fact, with a more solid material, such as wood, plastic, or even plaster, these cells could be made with a peg-and-hole arrangement to keep the parts together. If a good strong paper with a smooth finish is used, the parts may also be made to adhere with a thin rubber cement.

The Small Stellated Dodecahedron

To make a model of the small stellated dodecahedron uniform in construction with the models previously described, all you need for a net is an isosceles triangle with base angles 72 degrees and vertex angle 36 degrees. This is triangle number 1 in Figure 20. With a needle, prick through the vertices of this triangle placed as a net on top of six sheets of paper, each of a different color. You will need to repeat this pattern ten times, giving you ten triangles of each of the six colors—sixty triangles in all. Score the lines and trim the tabs as described before. You will find that the more acute angles must also have more acute trimming. This is best done after the tabs have been folded; if it is done before, folding becomes more difficult. (See Figure 21 for suggestions.) Next, five of these triangles are cemented together as in Figure 22. Then the final edges are joined to form a

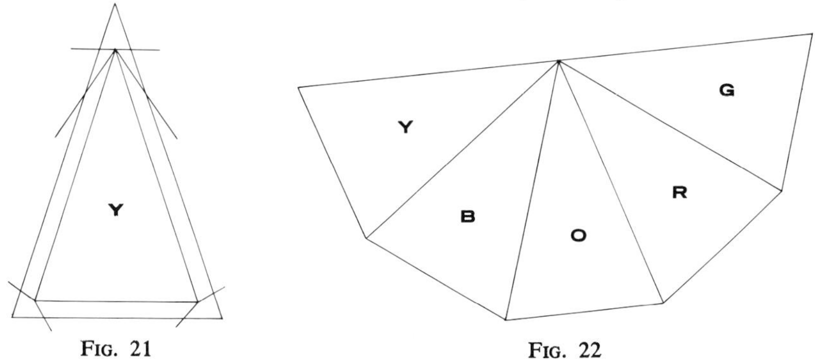

FIG. 21 FIG. 22

pentahedral angle—that is, a pentagonal pyramid without a base. You will find it easy to get the correct color arrangements by following the color scheme given below:

(0)	Y	B	O	R	G
(1)	W	G	O	R	B
(2)	W	Y	R	G	O
(3)	W	B	G	Y	R
(4)	W	O	Y	B	G
(5)	W	R	B	O	Y

Note that only six vertices are given here. The other six are made in a reverse order of arrangement, found by reading the table above from right to left, rather than from left to right as you did for the first six vertices. You must of course proceed in a systematic fashion, cementing the parts with all triangles pointing, say, away from you and working from left to right. In doing the second set of six you must work in the same manner, although you are reading the color table backwards. You will also find it more interesting if you cement the vertices together as you complete them. (Figure 23*a* shows how this is done to get the colors correct.) Each of the six vertices in the second set is placed diametrically opposite to its counterpart in the completed model.

As the models now come to be more intricate, with both convex and concave parts, a further hint about construction may be useful. The first parts are usually easy to handle while you can work on the interior where the tabs are being joined, using clamps as before. (As you proceed you will notice that the interior actually begins to look just as beautiful as the exterior, but of course this will all be hidden in the completed model. It just happens to be a fact that adds to your interest and enthusiasm for making models and thus lightens the monotony of the work of repeating so many parts.) It might seem that considerable skill or patience is needed to get the last vertex or part cemented in its proper place. But the secret here is to cement only one flap or tab first, let it set firmly, and then close the final opening, as described before. The double tabs make it easy to get the parts to adhere without clamps, since the model has sufficient form by this time to exert its own pressure.

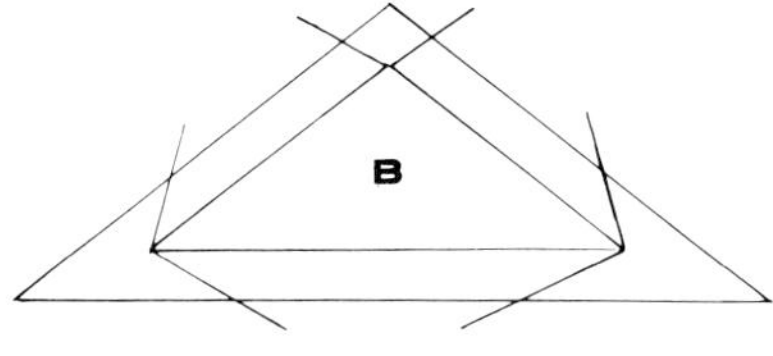

FIG. 23

By this method all the solids now being described can be constructed so that they are completely hollow inside. A basic solid could of course be used—for example, beginning with a dodecahedron, you could cement the vertex parts, twelve pentagonal pyramids, one onto each face, and thus obtain a small stellated dodecahedron. But it will be found that the final product will betray its construction when closely examined. In some models you may gain better rigidity by this method; in fact, the construction of the small stellated dodecahedron as described here is one in which the model is not technically rigid. But if the cementing is carefully done along the full length of each edge and a final drop of cement is added to both the acute and the obtuse or concave vertices in the completed model, you will find the result satisfactory.

Again, you may wish to economize on the number of separate parts used in a model. You might make one net, for example, in which the five triangles are all of one piece, giving a vertex part that is all of one color. You will find, however, that you have sacrificed beauty for economy. Once more, the rule here is that you get results proportionate to the efforts you put forth. These general comments apply equally to the models now to be discussed.

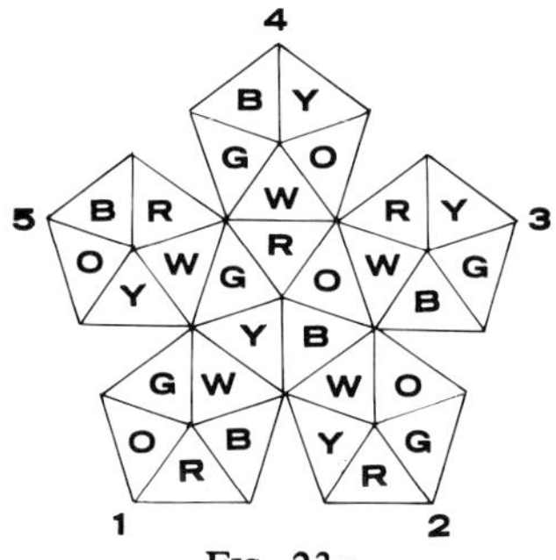

FIG. 23*a*

The Great Dodecahedron

The second stellation of the dodecahedron is known as the great dodecahedron. It may be described as a solid composed of twelve intersecting pentagons. When it is appropriately made in six colors, it readily gives the appearance of a solid star embossed on a pentagon plane; but each such star shares each of its arms with an adjacent star. The net is simply an isosceles triangle, this time with base angles of 36 degrees and a vertex angle of 108 degrees. This is triangle 2 in Figure 20. Ten sheets of each of the six colors are needed. (See Figure 23 for a suggestion about how to trim the parts. The color arrangement is shown in Figure 24.) Again only half the model is shown. A simple method of construction is to cement three triangles in the form of a trihedral dimple, as follows:

(1)	Y W G	(6)	G O Y
(2)	B W Y	(7)	Y R B
(3)	O W B	(8)	B G O
(4)	R W O	(9)	O Y R
(5)	G W R	(10)	R B G

Once this half has been constructed, the color pattern itself is evident enough to enable you to continue without further difficulty. The other ten dimples are the usual counterparts. In approaching the last pieces you must take care to have one trihedral dimple left as the last part to be cemented. Let the cement set firmly along one edge of this part, then apply cement to the other two edges and close the triangular hole, with the trihedral dimple serving as the lid. In this way the longer and acute part of the wedge-shaped star arms can easily be pinched together with the fingers.

The Great Stellated Dodecahedron

The third and final stellation of the dodecahedron is called the great stellated dodecahedron. This solid makes a lovely Christmas decoration, and is often seen as such in store windows and commercial displays during the holiday season. It can be, and often is, made by adding triangular pyramids to an icosahedron base; but, as mentioned before, this method will be found to lack something of the precision

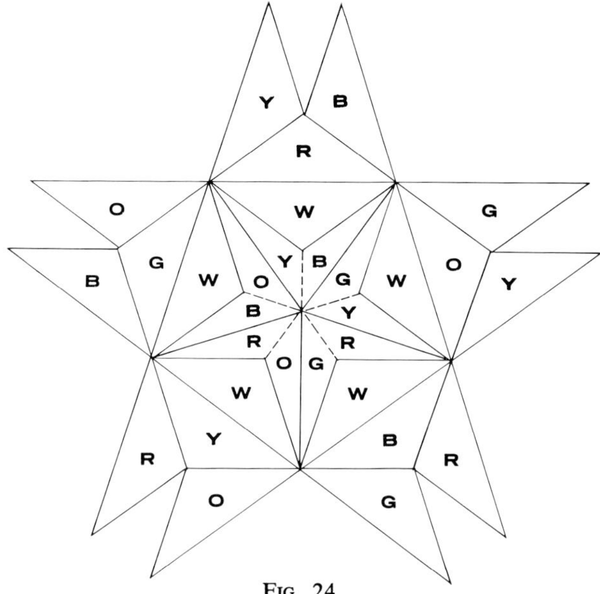

FIG. 24

and beauty that can be achieved in a completely hollow model in six colors. You may perhaps be surprised that the icosahedron is used as a base for a stellated form of the dodecahedron. But then there are many surprises in the world of polyhedra, and curiosity about the reasons for them may spur you to further study on your own. (See the Bibliography for related works.)

The net for the great stellated dodecahedron is once more the isosceles triangle with 72 degrees for two angles and 36 degrees for the third. It is triangle 3 in Figure 20. Ten sheets of each of the six colors will suffice. In this model, however, only three triangles are used to form a trihedral vertex, and twenty such vertices are needed for the complete model. One such vertex is shown in Figure 25. The color arrangement for ten vertices is as follows:

(1) Y G B　　(6) W G B
(2) B Y O　　(7) W Y O
(3) O B R　　(8) W B R
(4) R O G　　(9) W O G
(5) G R Y　　(10) W R Y

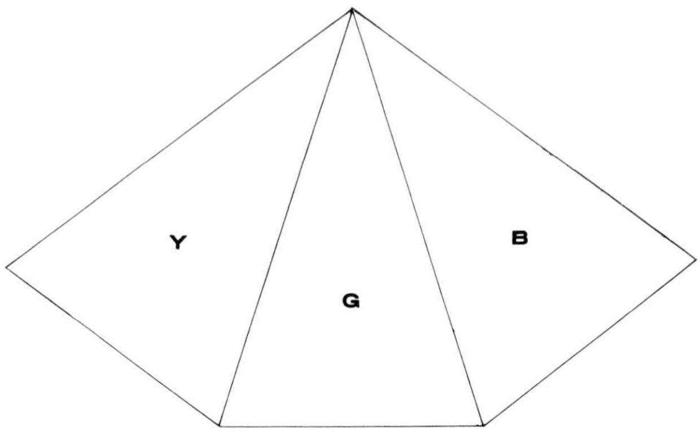

FIG. 25

The first five vertices are joined in a ring with the bottom edges forming a pentagon. Then the next five vertices are added to each edge of this pentagon, so that the white edge of (6) is cemented to the yellow of (1), and so forth. The next ten vertices have colors in reverse order of rotation and are placed diametrically opposite to their counterparts. If you work systematically, as before, you will not find this difficult. In fact, the colors will help you if you remember that each triangle is an arm of a five-pointed star and that you want each star to have five arms of the same color. When you have completed these three according to the instructions given here, you will also notice that planes that are parallel to one another are the same color.

Nothing has been said about the size of these polyhedra. Perhaps experiment will suggest appropriate measures, depending on how and where you wish to use the models. Different types of paper or cardboard also may be used: shining gold, silver, green, or red would give breathtaking results in a model hung by thread and illuminated with various lights!

A complete set of four dodecahedra—the original dodecahedron and its three stellations—all using the same six colors, and all built to display their exact relationship in size to one another, makes a most attractive classroom display. And when you have made these and examined them at close range, you will better understand the principle of stellation.

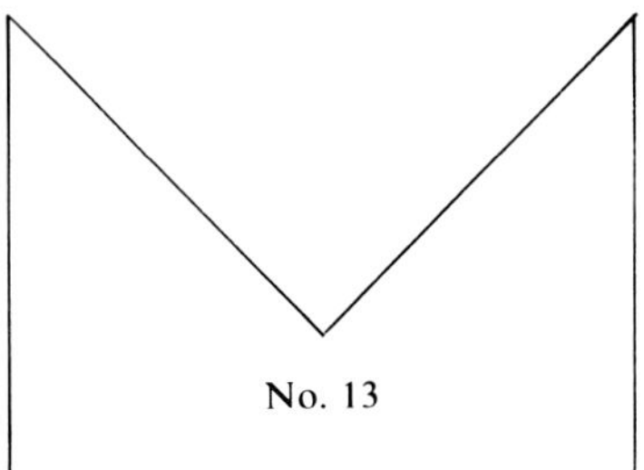

No. 13

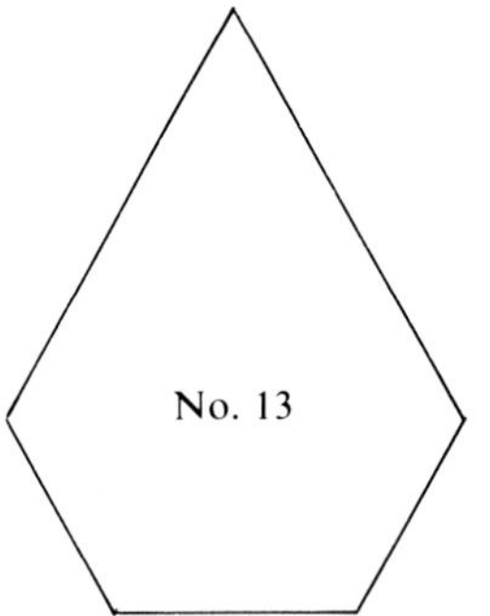

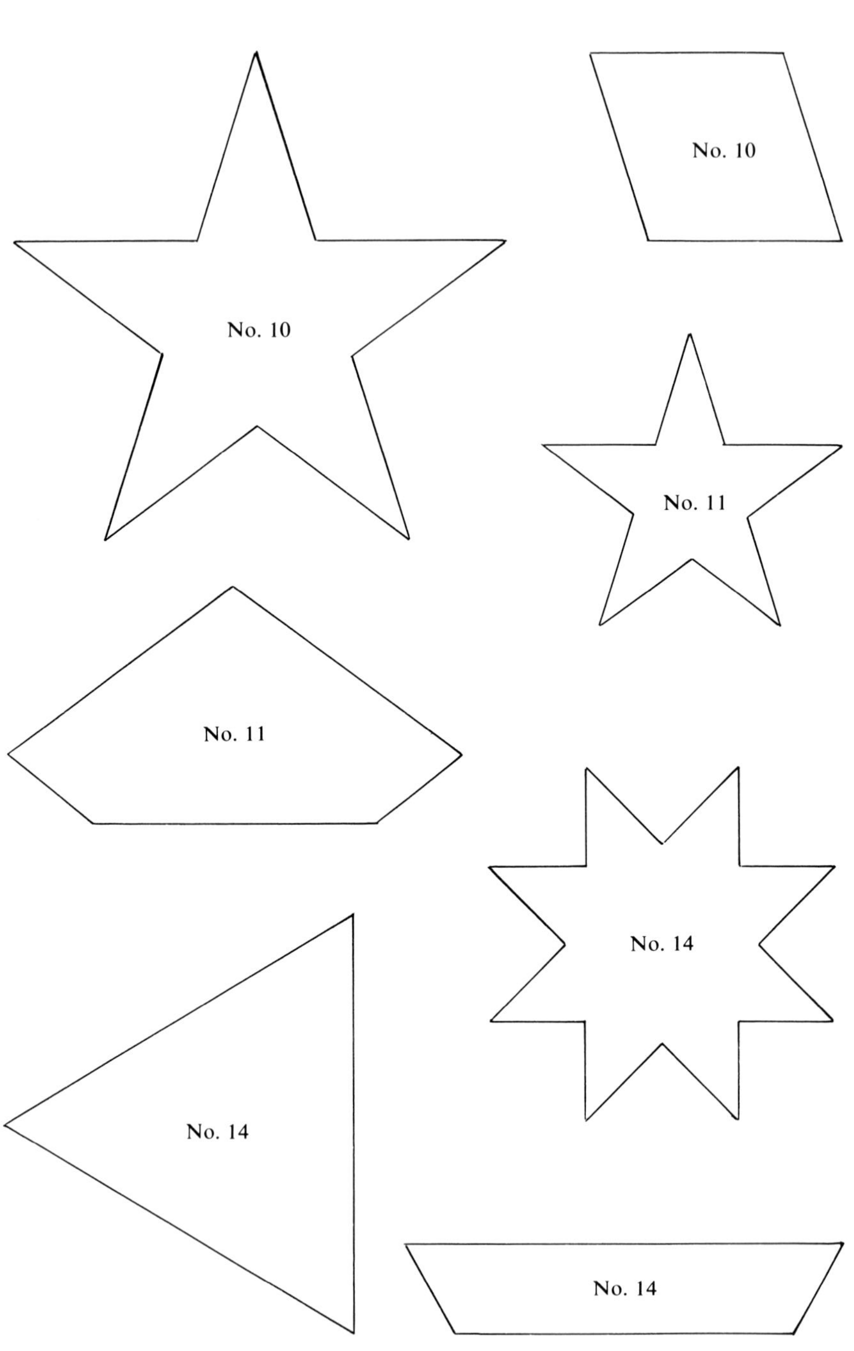
No. 10
No. 10
No. 11
No. 11
No. 14
No. 14
No. 14

SOME OTHER UNIFORM POLYHEDRA

(See pp. 28-40. Models are numbered as shown on pp. 28-29.)

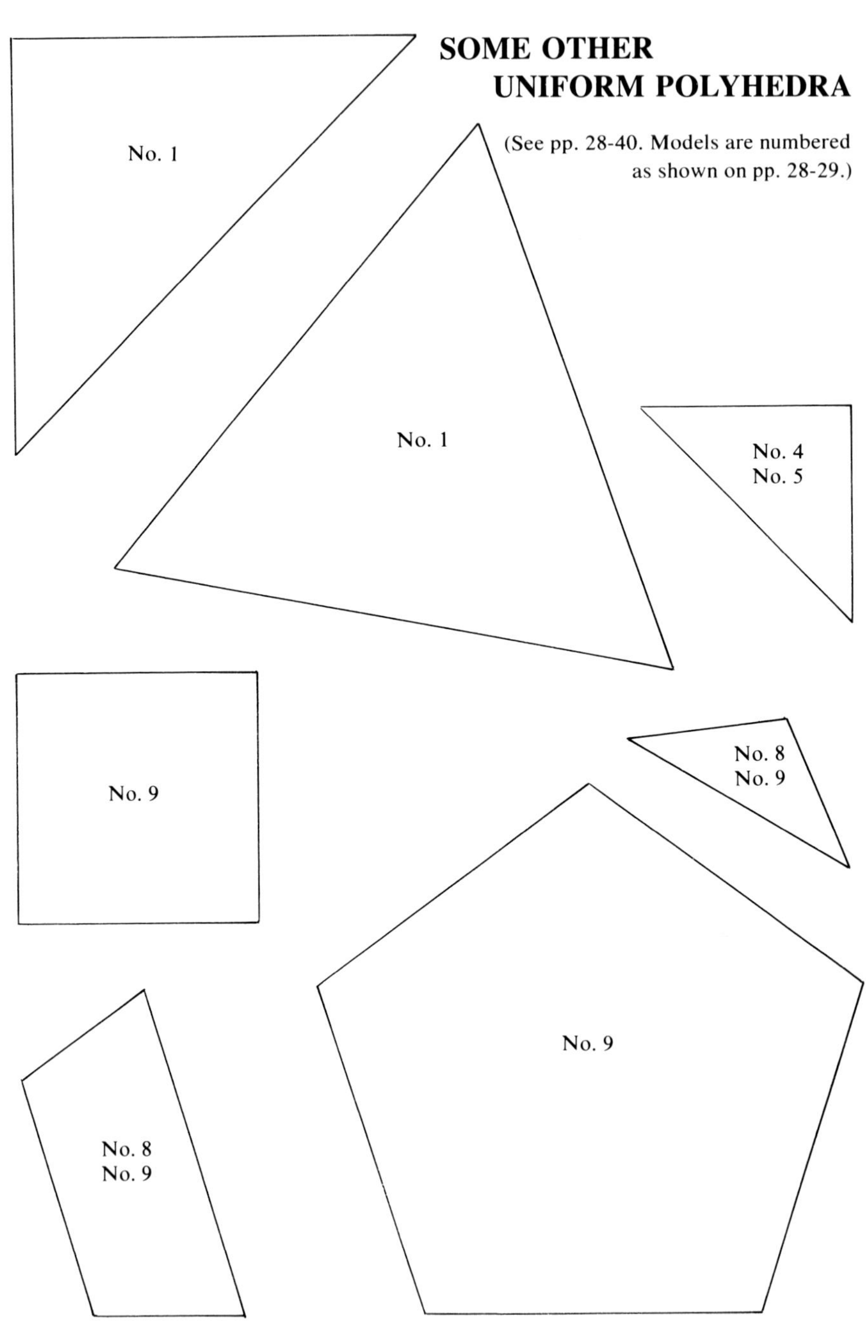

OTHER STELLATIONS OR COMPOUNDS

(See pp. 21-27.)

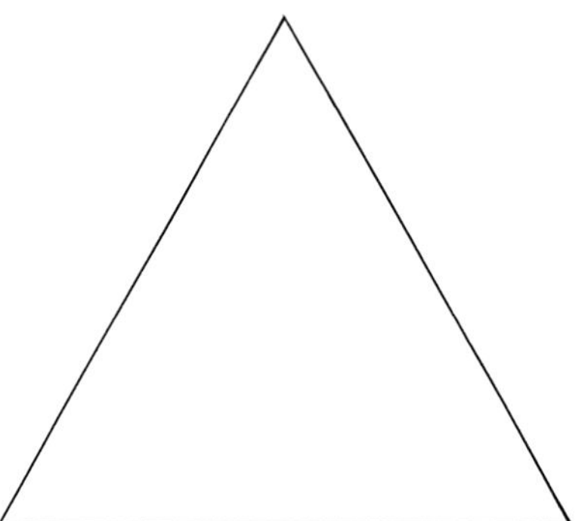

COMPOUND OF TWO TETRAHEDRA
(KEPLER'S STELLA OCTANGULA)

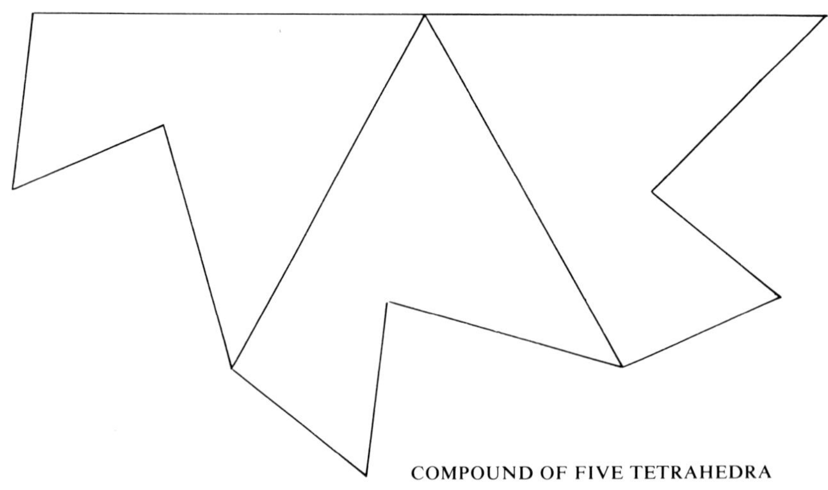

COMPOUND OF FIVE TETRAHEDRA

THE FOUR KEPLER-POINSOT SOLIDS

(See pp. 11-20.)

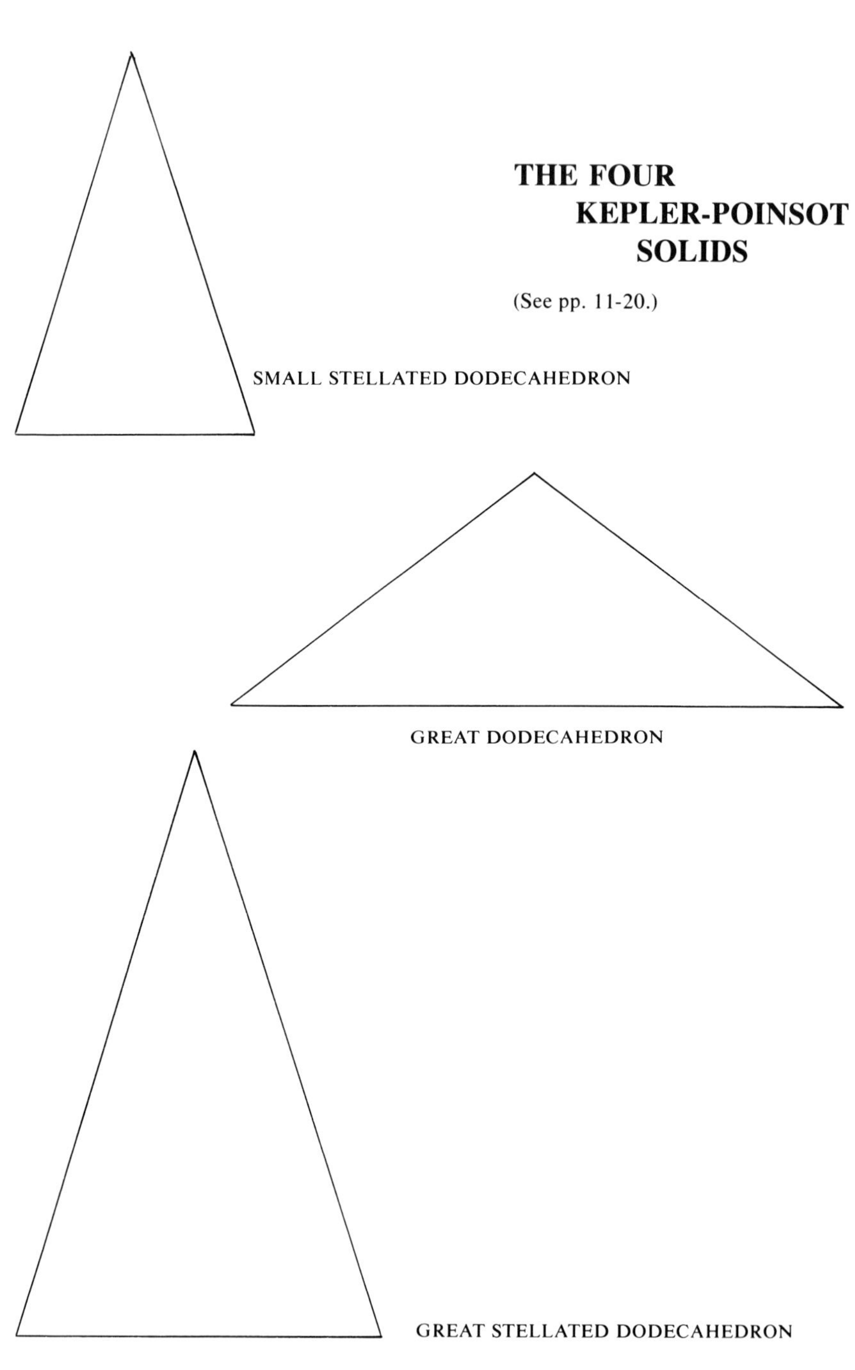

THE THIRTEEN ARCHIMEDEAN SOLIDS

(See pp. 6-8.)

THE FIVE PLATONIC SOLIDS

(See pp. 2-4.)

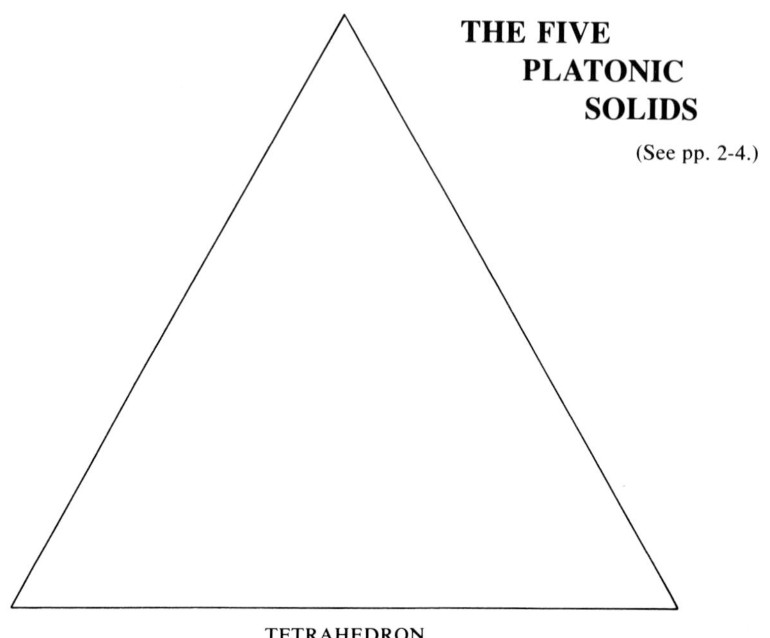

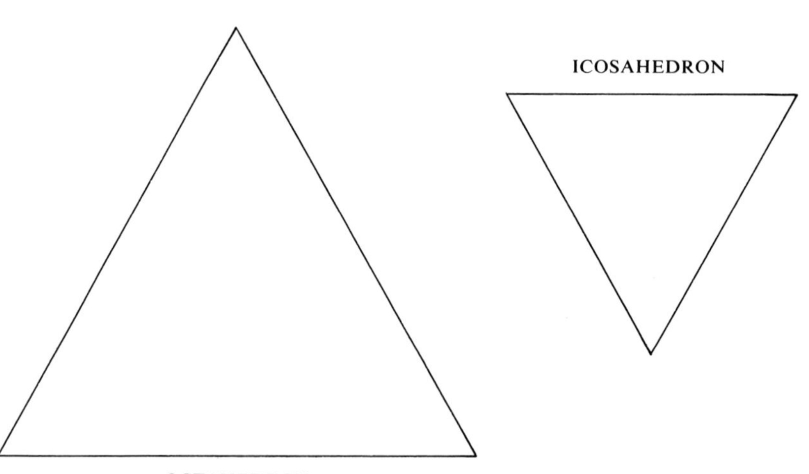

HEXAHEDRON (CUBE)

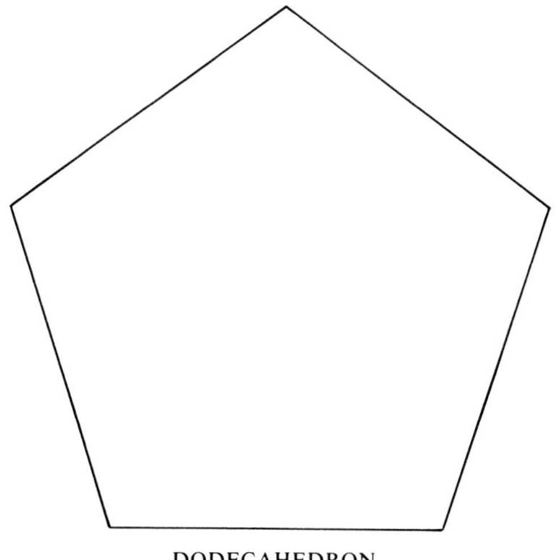

DODECAHEDRON

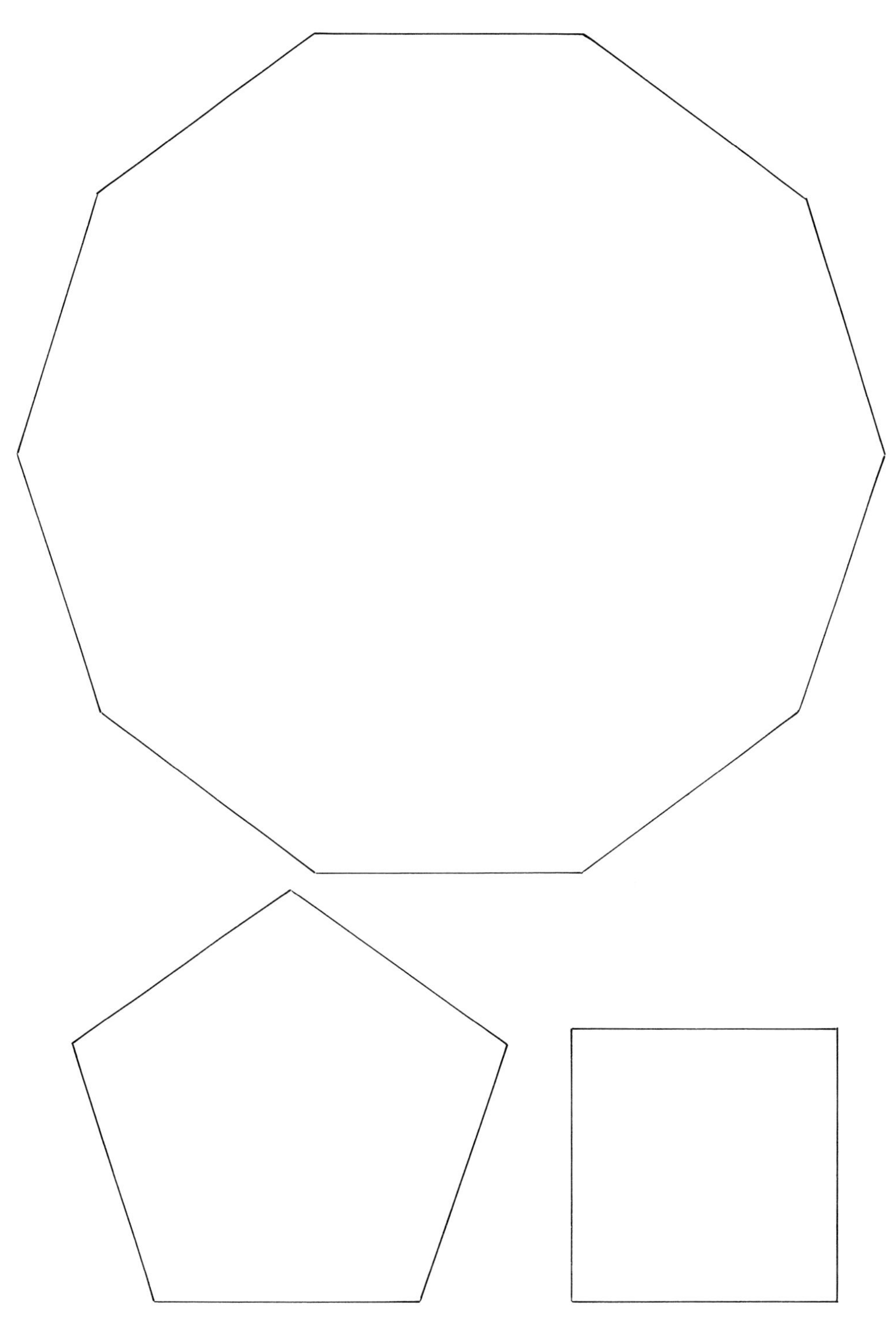

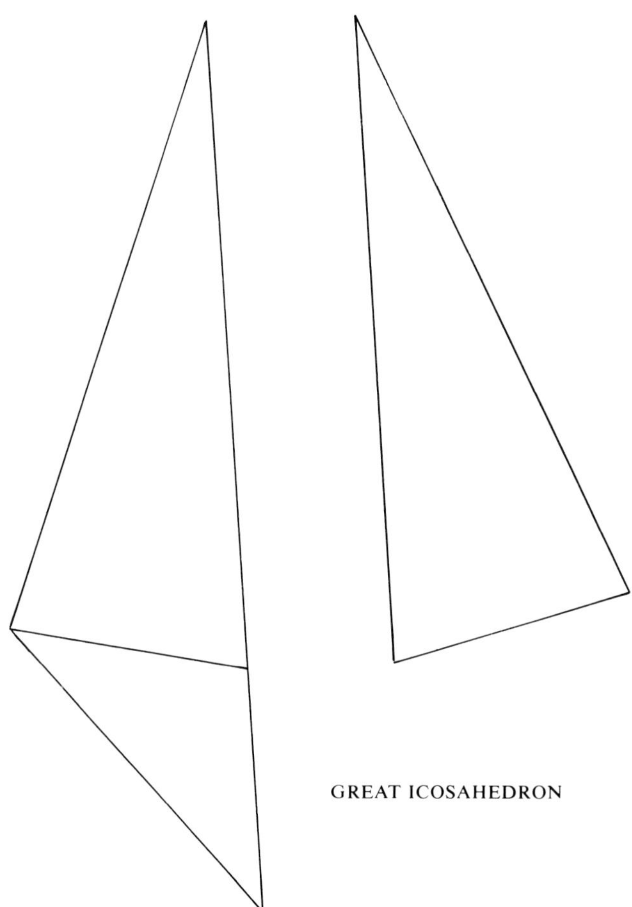

GREAT ICOSAHEDRON

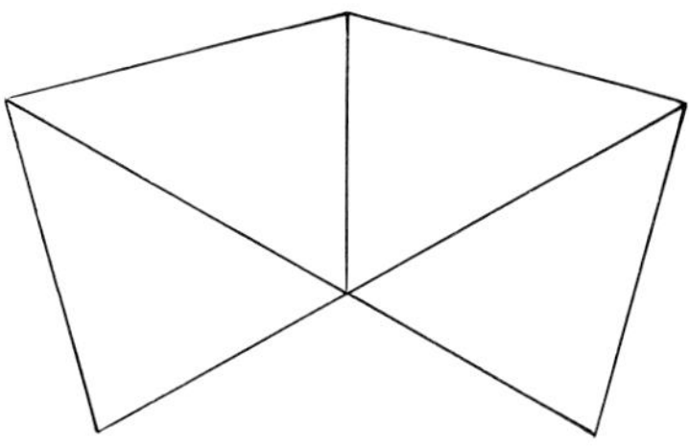

COMPOUND OF FIVE OCTAHEDRA

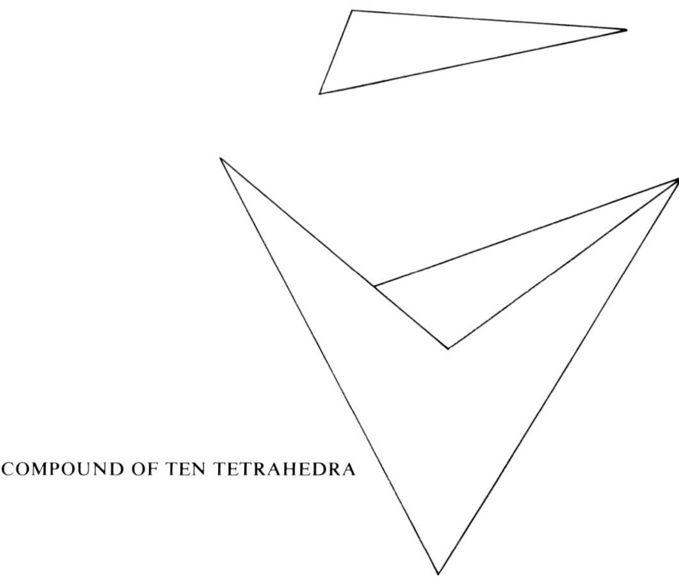

COMPOUND OF TEN TETRAHEDRA

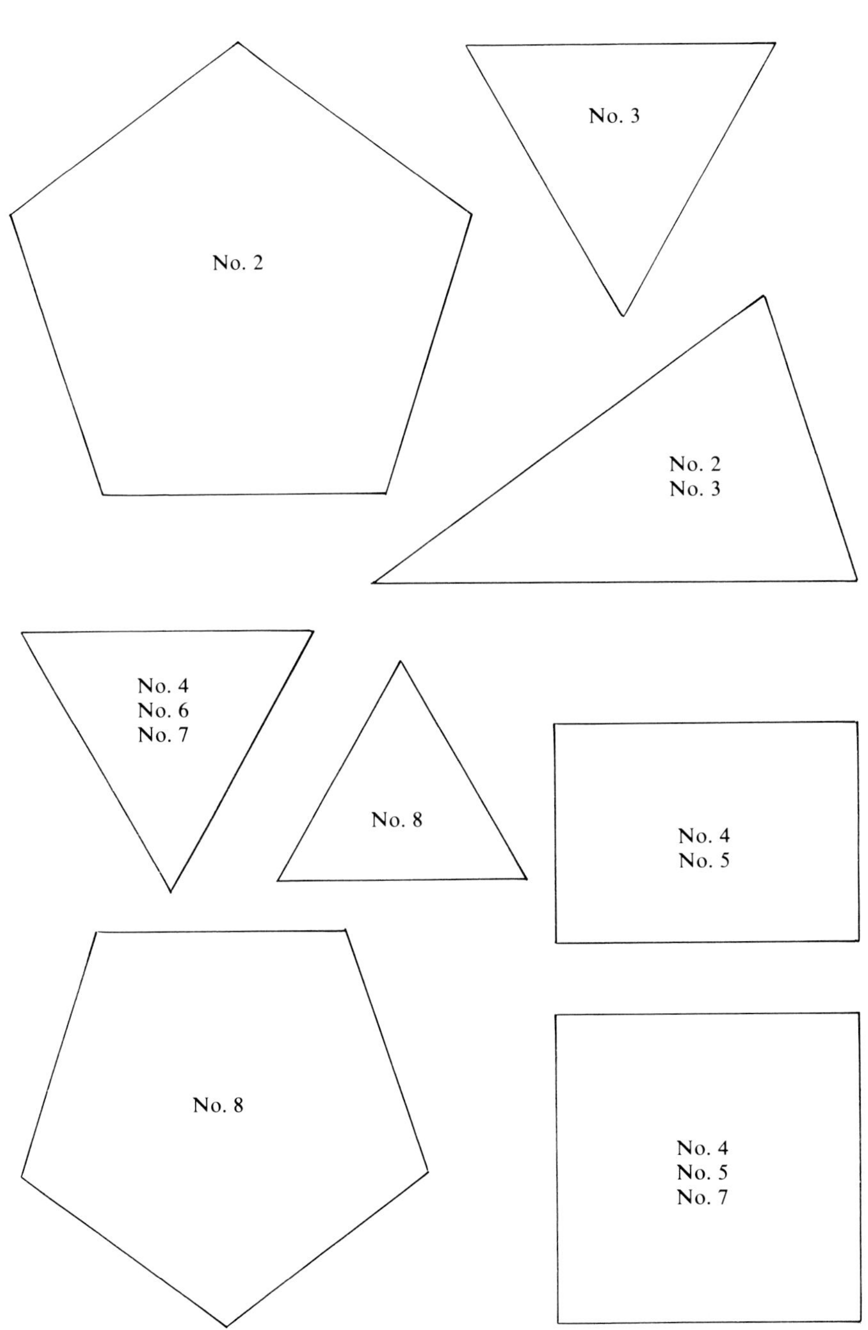
No. 2
No. 3
No. 2
No. 3
No. 4
No. 6
No. 7
No. 8
No. 4
No. 5
No. 8
No. 4
No. 5
No. 7

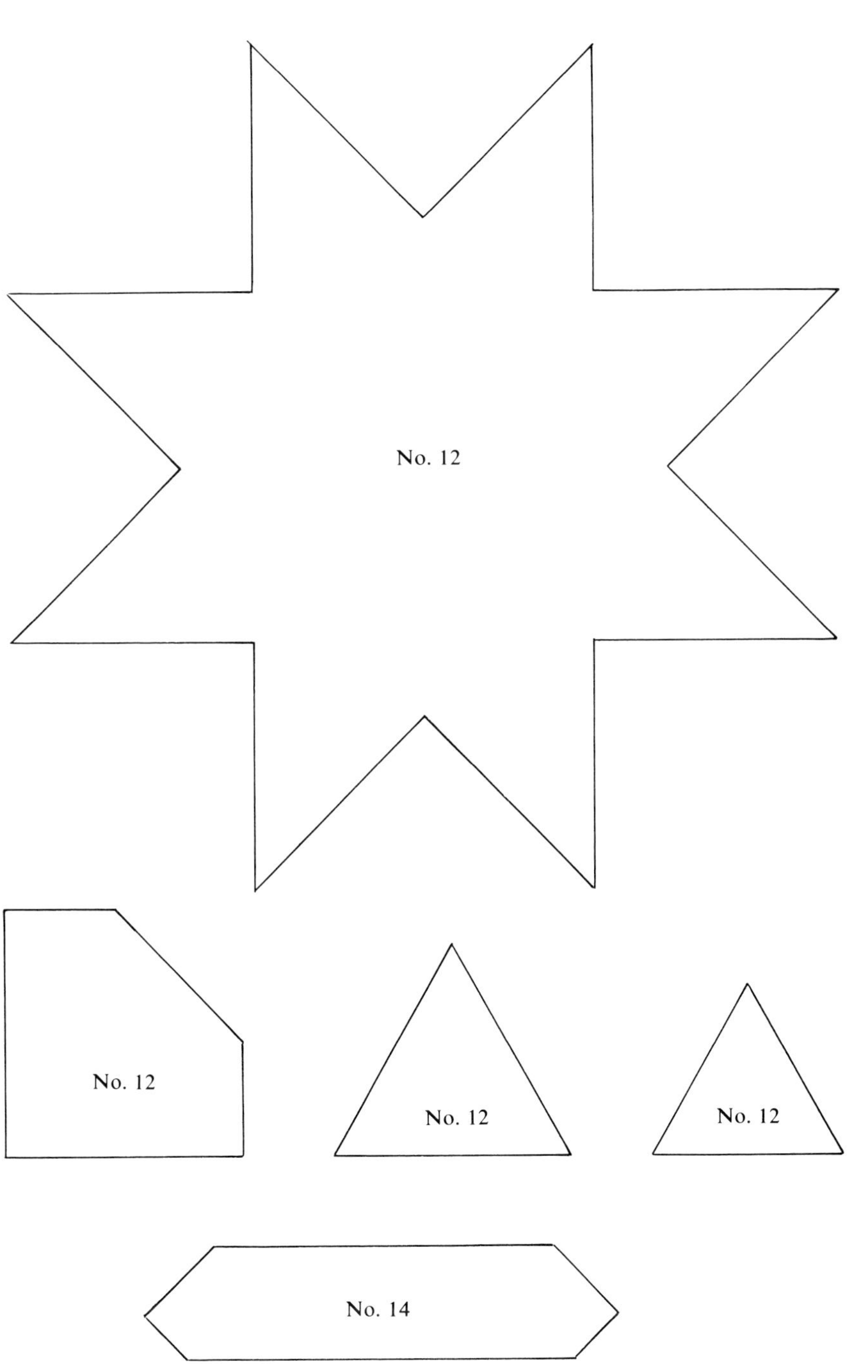
No. 12
No. 12
No. 12
No. 12
No. 14

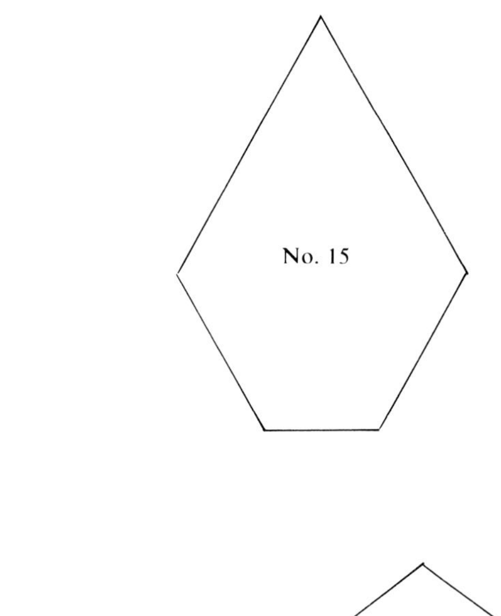
No. 15

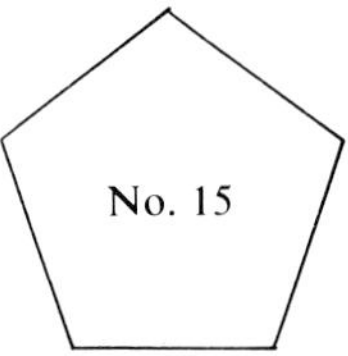
No. 15

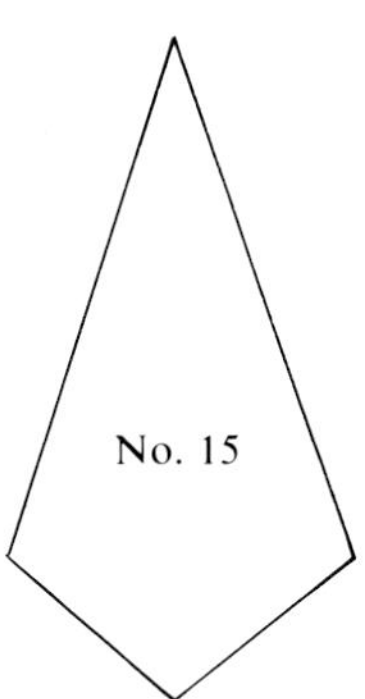
No. 15

The Great Icosahedron

But of all the polyhedra so far described perhaps the most beautiful and attractive is the great icosahedron itself, which is the fourth of the Kepler-Poinsot solids. The nets are very simple (see Figure 26). The color arrangement described here gives a final beauty that cannot but appeal to anyone who sees or handles a model. And, fortunately enough, it is not at all difficult to make—not as hard as some of the compounds mentioned below. It does require the patience and the time to prick, cut, score, and trim one hundred twenty individual pieces of cardboard for a complete model. But it is well worth the trouble and the effort.

First staple together five sheets of colored paper, one sheet of each of the five colors. Prick out the patterns to get copies of the net, score, and trim as usual. The dotted line of Figure 26 must be scored on the reverse side, since the small isosceles triangle is folded up. The tabs, as usual, are folded down. You are now ready to cement the parts together. Follow the paired arrangement of colors given below, cementing the pairs first:

(0)	YG	BY	OB	RO	GR
(1)	BG	YB	RY	OR	GO
(2)	OY	BO	GB	RG	YR
(3)	RB	OR	YO	GY	BG
(4)	GO	RG	BR	YB	OY
(5)	YR	GY	OG	BO	RB

Then five pairs make the fanlike form shown in Figure 27. When the remaining edges are brought together, a vertex is completed. The fold should be down between each member of the pairs given above, and up between the pairs. The smaller isosceles triangles should then be cemented in their respective places to form a pentagonal dimple, from which the vertex rises. Twelve vertices are required; as before, six are counterparts of the other six. You will automatically find the counterparts by reading the color table in reverse, provided you continue the same systematic handling of parts for all vertices. The vertices are joined as shown in Figure 33, where the colors indicated are the colors of respective dimples.

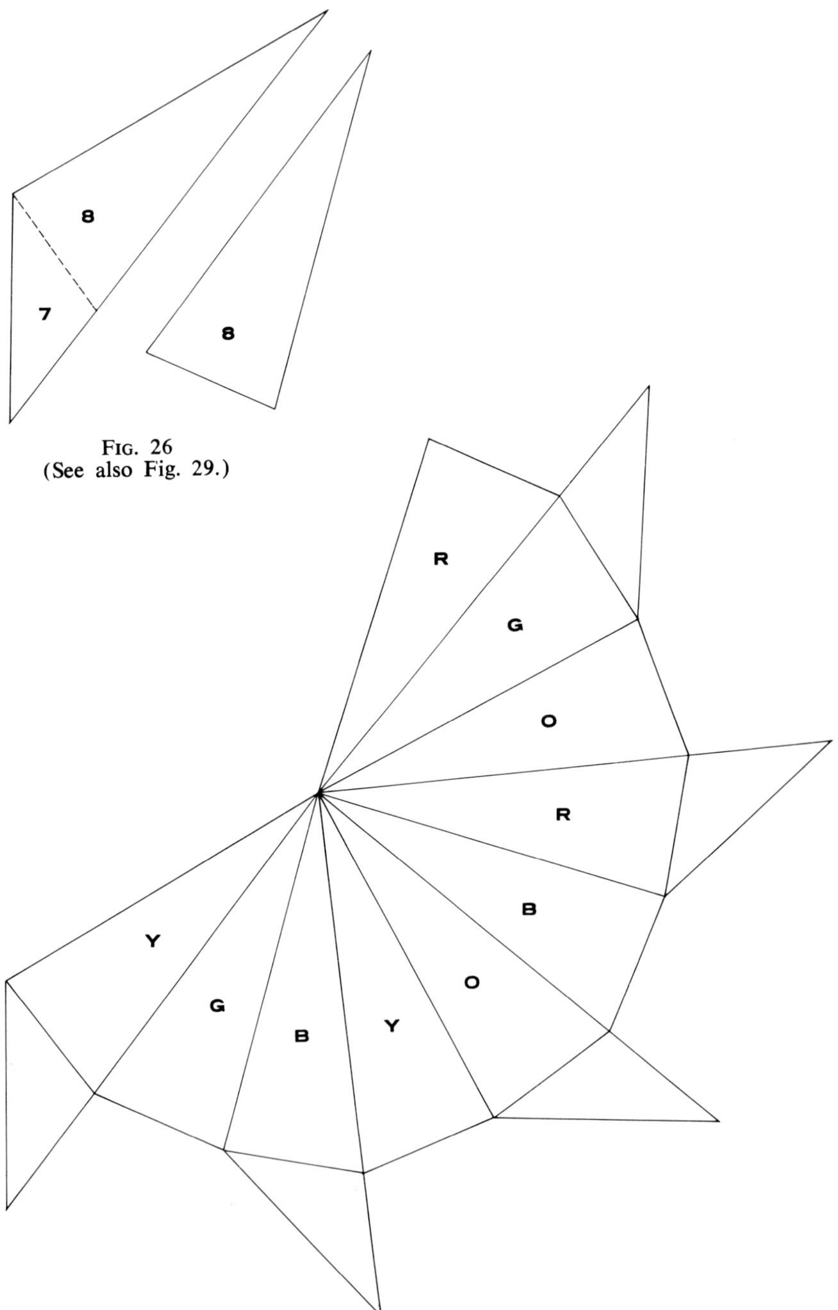

FIG. 26
(See also Fig. 29.)

FIG. 27

Other Stellations or Compounds

The principle of stellation can be applied to only two other Platonic solids: the octahedron and the icosahedron. If you have studied the principle as it applies to the dodecahedron in the three stellated forms described above, you may also understand very readily why the cube and the tetrahedron cannot be stellated. It is interesting that the octahedron has only one stellation. This is the eight-pointed star—or *stella octangula*, as Kepler called it—which actually turns out to be a compound of two tetrahedra. It is even more interesting to find among the stellations of the icosahedron other compounds; but more about these later. None of these compounds is classified with the uniform

COMPOUND OF TWO TETRAHEDRA (KEPLER'S "STELLA OCTANGULA")

COMPOUND OF FIVE OCTAHEDRA

COMPOUND OF FIVE TETRAHEDRA

COMPOUND OF TEN TETRAHEDRA

OTHER STELLATIONS OR COMPOUNDS

Model at top left is a stellation of the octahedron. The rest are stellations of the icosahedron.

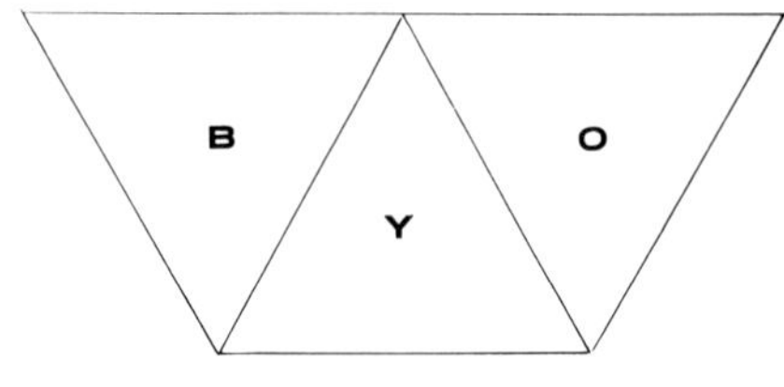

FIG. 28

polyhedra, precisely because they are compounds; specifically, they are intersecting polyhedra or interpenetrating polyhedra, not intersecting polygons. Nevertheless these compounds are true stellations, and they do make pleasing models. It is for this reason that a description of them is included in this monograph.

To make a model of the *stella octangula,* all you need for a net is an equilateral triangle. Since there are eight trihedral vertices, the color arrangement may be as follows:

(1) B Y O
(2) O R Y
(3) B Y R
(4) R O B

One vertex is shown in Figure 28. The other four vertices are the reverse order of counterparts of this, and you should find no difficulty in making the proper assembly if you remember that in this arrangement of colors each of the four sides of each tetrahedron is a different color, but each pair of parallel planes of the polyhedron is the same color.

The icosahedron has some very interesting stellations. Including the compounds mentioned above, the total enumeration comes to fifty-nine, if one follows the complete analysis of the problem given by Coxeter.[4] It is indeed surprising that a compound of five octahedra, a compound of five tetrahedra, and a compound of ten tetrahedra appear among the stellations of the icosahedron. Such a fact would have delighted the mind of Plato.

These compounds make very attractive models. To understand how the nets are obtained, it is necessary to know something about the

8 8
5 6 6 5
4 4
7 3 3 7
1 1
3 2 2 3
4 4
O
5 1 1 5
6 6
8 8
1 1
6 6
2
5 5
4 4
3 3
8 8
7

FIG. 29

1 1
1 Y Y 1
Y Y

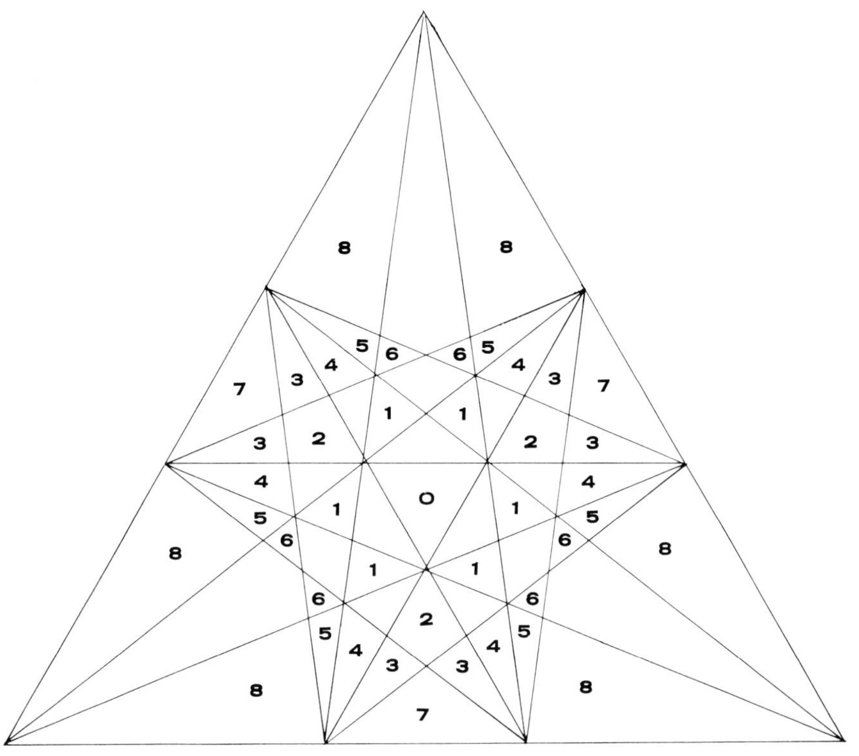

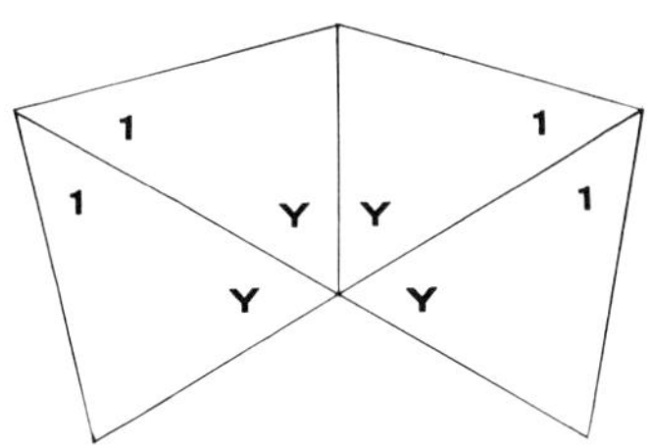

FIG. 30

stellation pattern for the icosahedron. This is analogous to the dodecahedral pattern of Figure 20. The icosahedral pattern is shown in Figure 29. Actually, the innermost equilateral triangle (numbered 0) is one of the faces of the icosahedron; and the outermost equilateral triangle is one of the facial planes of the great icosahedron, the fourth of the Kepler-Poinsot solids. If each side of this large triangle is divided by two points according to the "golden section"—a linear section, discussed in Euclid's *Elements,* that is approximately 1:0.618—the pattern is quickly and easily drawn. The numbering will show what parts are used for each net. (The nets are given in Figures 30, 32, and 34.) All of these compounds can be made by the methods described above—that is, by using parts with tabs left for cementing the pieces together and constructing the models so that they are completely hollow inside. The color patterns are such that in the case of the compounds of five octahedra and of five tetrahedra each solid is of one color. In the compound of ten tetrahedra each two tetrahedra that share facial planes also share a color.

To construct the compound of five octahedra, make thirty copies of the net in Figure 30, six of each of the five colors. First assemble the vertices as though they were small pyramids without their rhombic bases. Then follow the color pattern shown in Figure 31, where each rhomb is a vertex. This shows a ring of five vertices at the center. Between the extending arms of this ring a second set of five vertices is cemented, but their orientation is such that the short slant edge of each pyramidal vertex continues on a line with the grooved edge between vertices of the central ring. You may find this a bit puzzling; but if you remember to keep the basic octahedral shapes in mind, you will see them begin to develop, and the color will then help you proceed correctly. The color pattern of Figure 33 now begins to appear. By comparing the numbering of Figure 31 with Figure 33, you will see this. This hollow model is not completely rigid, but it will be satisfactory nevertheless.

For a model of the compound of five tetrahedra all you need is twenty copies of the net in Figure 32, four of each of the five colors. First make trihedral vertices with the bottom edges looking rather jagged. If you begin by making a ring of five vertices cemented together with the edge marked *AB* of one adhering to the identical edge

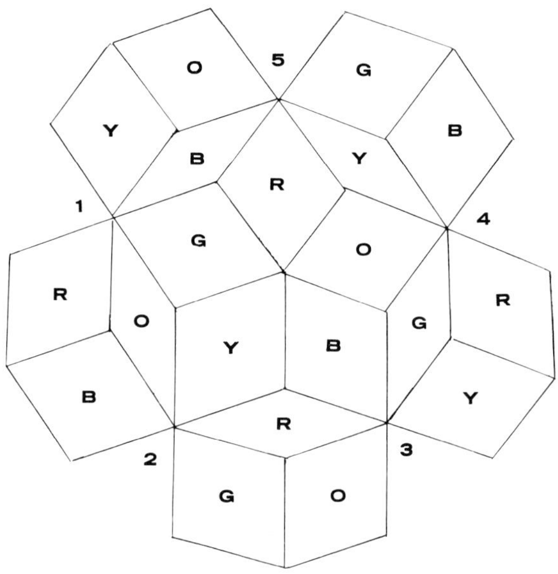

FIG. 31

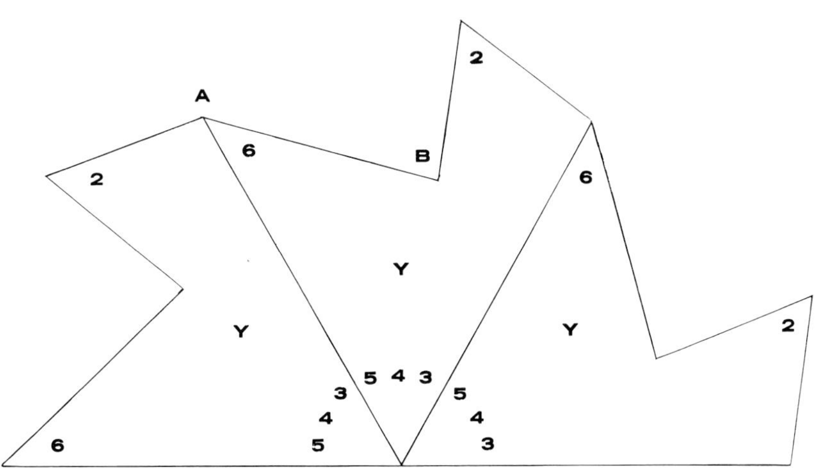

FIG. 32

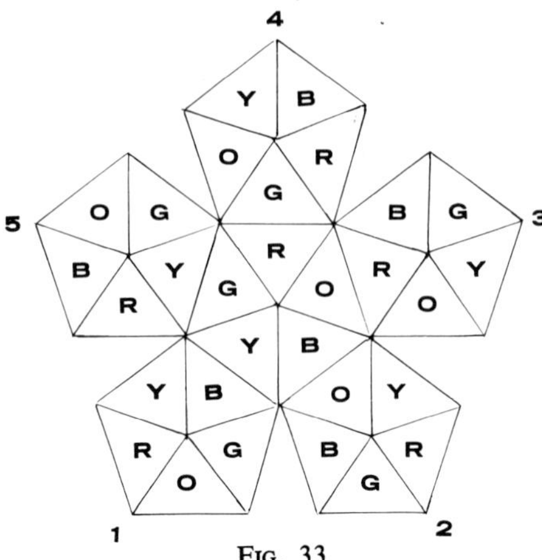

FIG. 33

of the other, you will find the points numbered 2 forming a dimple in the center of the ring. Once you have built this much of the model, the other vertices will easily find their places according to the color scheme, making each tetrahedron entirely of one color. This is perhaps the most difficult model to construct because of all the jagged edges. The points numbered 2 fit into three different and adjacent dimples. The secret here is to worry about only one edge at a time. Always begin with the edge *AB*. Once it is cemented, let it set firmly, and then give your attention to the other edges. You may find that the last vertex will call for considerable skill, not to mention patience; but it can be done. Here you have a real challenge. This model is rigid, both technically and practically; and it is also aesthetically pleasing. (Folding the parts up instead of down gives the model a reverse twist —a twist that, by the way, will not be noticed by most people.)

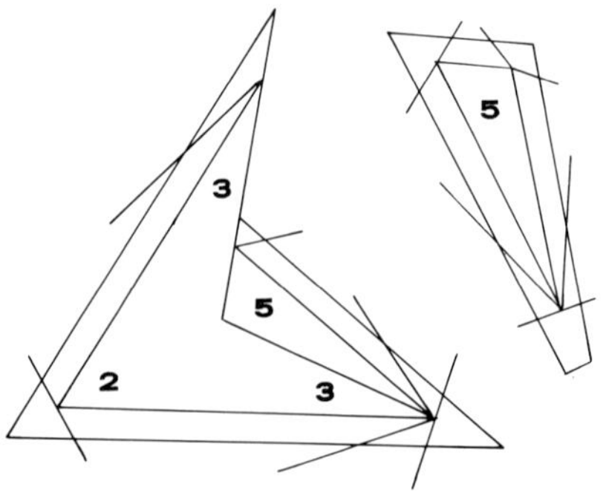

FIG. 34

For a model of the compound of ten tetrahedra, begin by making sixty copies of each of the two parts shown in Figure 34—that is, twelve copies of each of the five colors. Suggestions for trimming the tabs are shown with the nets for the sake of clarity. Note that the left arm of the net in Figure 34 is cut without leaving a tab on its right side, and that the cut is to be made clean into the center point between the arms. The triangle numbered 5 can then be folded down. Once five parts have been cemented together to form a pentagonal dimple, the other triangle numbered 5 can be cemented to the edge lacking a tab by following the color arrangement shown in Figure 35 (basically the same as that shown in Figure 33). The short dotted lines in Figure 35 indicate overlapping parts, but these parts are to be folded down so that the bases of the small triangles can be joined. This is done by folding up the tab of one to adhere to the undersurface of the other, which lacks a tab. This completes one part. Twelve of these parts are needed for a complete model, six of which are counterparts of the other six, as explained before. Some skill and patience is needed for this model also. But it can be done, and it makes a very pleasing polyhedron.

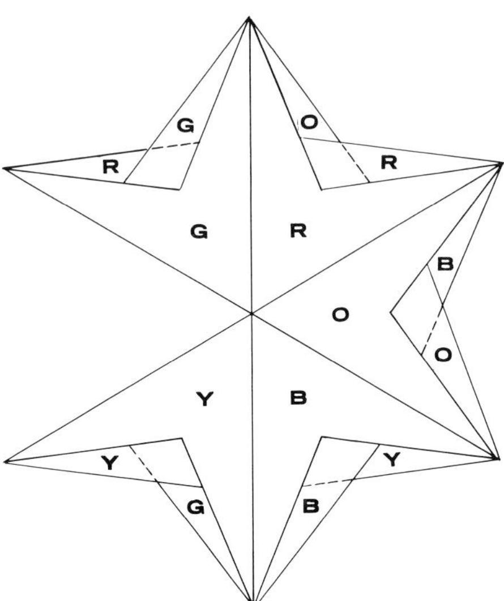

FIG. 35

Some Other Uniform Polyhedra

So far this monograph has described the five Platonic solids, the thirteen Archimedeans, the infinite set of prisms and antiprisms, a few miscellaneous irregular polyhedra, the four Kepler-Poinsot solids, and some compounds arising from the stellation process. All of these except the irregulars and the compounds are classified as uniform polyhedra. (A polyhedron is uniform when all its faces are regular polygons [including regular star polygons] and all its vertices are alike.) The question might now well be asked, Are there any more uniform polyhedra? The answer is, Yes, there are more—at least fifty-three others. Coxeter has published a brief history of all these

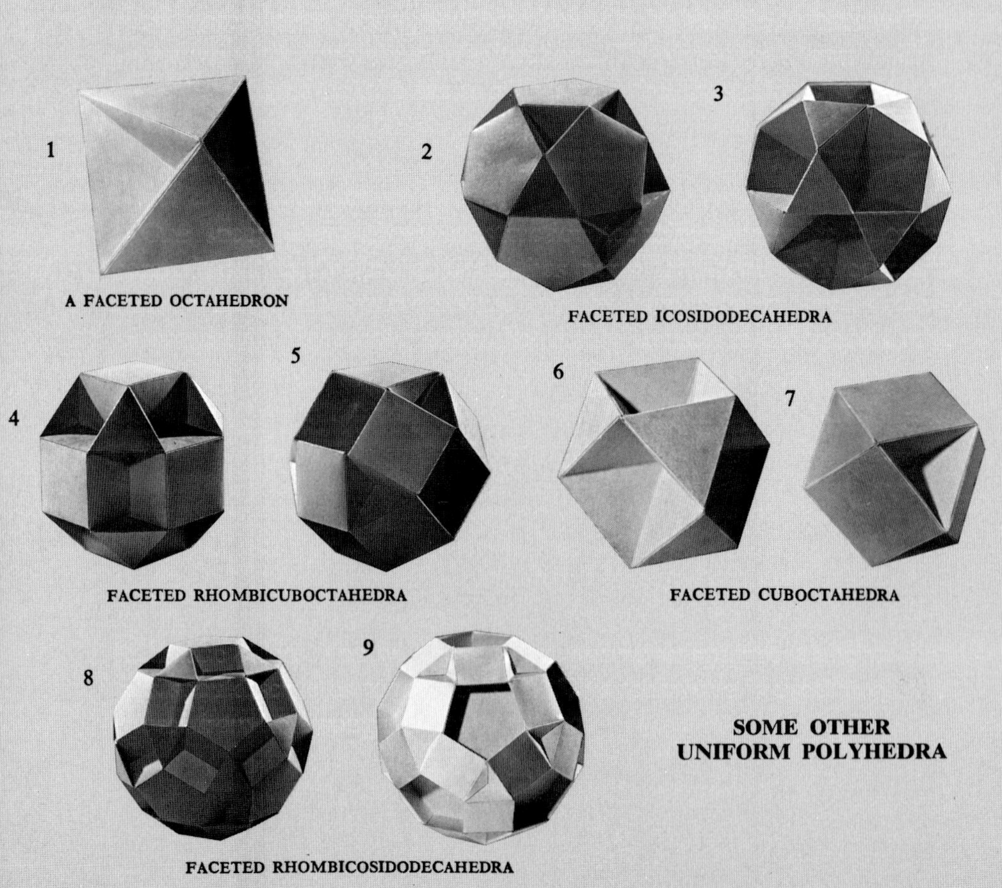

A FACETED OCTAHEDRON

FACETED ICOSIDODECAHEDRA

FACETED RHOMBICUBOCTAHEDRA

FACETED CUBOCTAHEDRA

FACETED RHOMBICOSIDODECAHEDRA

SOME OTHER UNIFORM POLYHEDRA

solids and a thorough investigation of the mathematics involved in their classification and discovery. In this book, entitled *Uniform Polyhedra,* he mentions the fact that Badoureau in 1881 made a systematic study of all thirteen Archimedean solids on the basis of their planes of symmetry and their vertices to discover polygons penetrating the interior parts of each solid. By this method he discovered thirty-seven uniform polyhedra not known before that time.[5] Brueckner published a classic work on polyhedra in 1900 in which many were beautifully illustrated in photographed models and in drawings.[6] About 1930 Coxeter himself and J. C. P. Miller discovered twelve other uniform polyhedra by investigating the Schwarz triangles on a spherical surface. Thus the total enumeration of uniform polyhedra

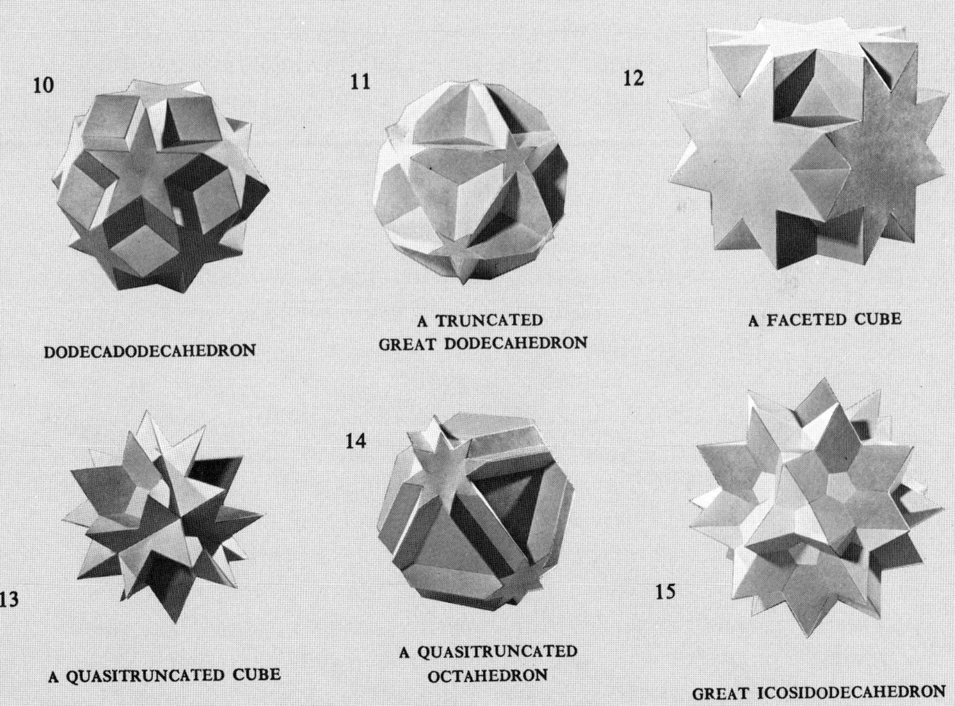

10 DODECADODECAHEDRON

11 A TRUNCATED GREAT DODECAHEDRON

12 A FACETED CUBE

13 A QUASITRUNCATED CUBE

14 A QUASITRUNCATED OCTAHEDRON

15 GREAT ICOSIDODECAHEDRON

NOTE.—A *faceted* polyhedron is one that may be derived from a convex polyhedron by the removal of solid pieces. A *stellated* polyhedron is one that may be derived from a convex polyhedron by the addition of solid pieces. *Truncation* is also a removal process. *Quasitruncation* is a combination of removal and addition.

given by Coxeter in 1954 comes to seventy-five, but (a most striking fact) he admits that a rigorous proof has still to be given that this enumeration is complete.[7]

As for the construction of models for these polyhedra, this monograph can present only a selection of some of the more simple ones, since some are actually most intricate. Coxeter gives sketch drawings, done by J. C. P. Miller, of all seventy-five and photographs of models in wire, made by M. S. Longuet-Higgins.[8] As for the most intricate of these, one may well suspect that only the original discoverer has ever had patience, zeal, or perseverance enough to make a model. The drawings alone are amazing enough.

For the purposes of this monograph the following fifteen models are described, as a good representative set of the simpler ones. When you have made them, you will have a good idea of why they are classified as uniform polyhedra. It should not be necessary to give a detailed description of the construction of these models. If you have been successful so far, you will without doubt be able to proceed on your own with a minimum of direction. The drawings shown in Figures 36 to 50 reveal the facial planes and the lines of intersection of planes, as well as the nets required for each model. If you compare each of these with the photograph of its model, you will understand the relationship of parts.

One special hint about construction techniques may be in order here. It is this: Where a model has parts that have only edges in common, a good way to construct it is by using a tongue-and-slot arrangement. This technique can frequently be used in the models now being described. For example, in Figure 36 you may begin by cementing the hypotenuse of each of the isosceles right triangles to an edge of the equilateral triangle. Then, instead of cementing these isosceles triangles to form a triangular pyramid with all tabs inside, turn one or two sets of tabs out, leaving the other set or sets turned in but not cemented. The tabs turned out may be cemented to form a tongue, which will later be inserted into the slot of another pyramidal part and cemented there. A little experimenting will soon make this technique clear. You will have to use your own judgment about what tabs should serve as tongue or slot. This technique is used in all the figures from 36 through 44.

As for color arrangements, no detailed description should be necessary. You will find the most satisfactory results are obtained if you make facial planes the same color. Try to follow the map-coloring principle. Parallel planes may always be of the same color. Thus in Figure 39 the octagons and squares that are parallel may be of the same color. In Figure 41 the pentagons and decagons that are parallel may be of the same color, since each decagon is on a plane below and parallel to a pentagon. In Figure 42 the pentagon is part of the same facial plane as the decagon, so it is of the same color. In Figure 45 each star is on a plane parallel to the pentagon below it, so a six-color arrangement works out very nicely. Figure 46 is easily recognizable as a truncated form of the great dodecahedron and may thus have the same color pattern. Each star can be of the same color as the decagon below it. In Figure 47 three colors may be used for the octagonal stars, and the same three colors will serve for the squares that lie below them. Then two other colors may be used for the triangles. These color arrangements are well worth the trouble it takes to get them done correctly, because they help so much to bring out the relationship between the intersecting regular and star polygons. Figure 48 is somewhat like Figure 47, but each small square is now coplanar with an octagonal star and thus should be of the same color as that star. In Figure 49 there are eight triangles and their related hexagons, which are actually coplanar. Then there are six octagons, each parallel pair being perpendicular to the other two parallel pairs. Finally the octagonal star is set high up over these octagonal planes. Two colors will serve for the triangle-hexagon set, so they can alternate; three other colors will then fill the needs of the rest of the planes, with parallel planes of the same color. Finally, Figure 50 is best made with a five-color arrangement for the triangles. It will be found that the color arrangement given above for Figure 33 will serve the purpose here. In fact, the construction can be easily done by making the pentahedral dimples each with a pentagon of a sixth color at the bottom. These dimples can then be joined by cementing the star arms to the edges of these dimples. A single net of three star arms joined at their bases will serve here, because all the star planes are of the same color, whereas each triangular plane has its own color. (See the following pages for Figures 36-50.)

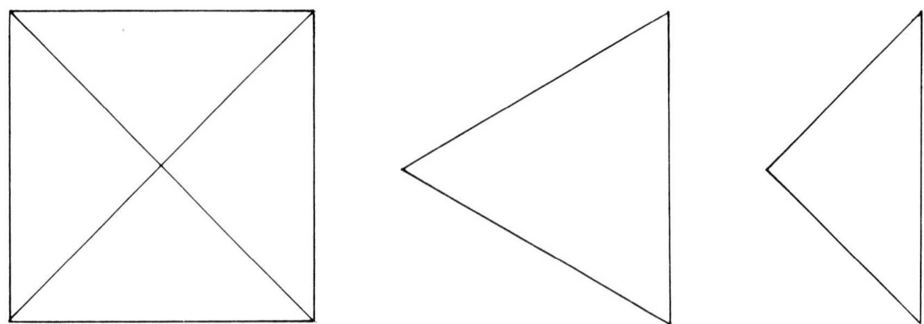

Fig. 36 (Model 1, p. 28)

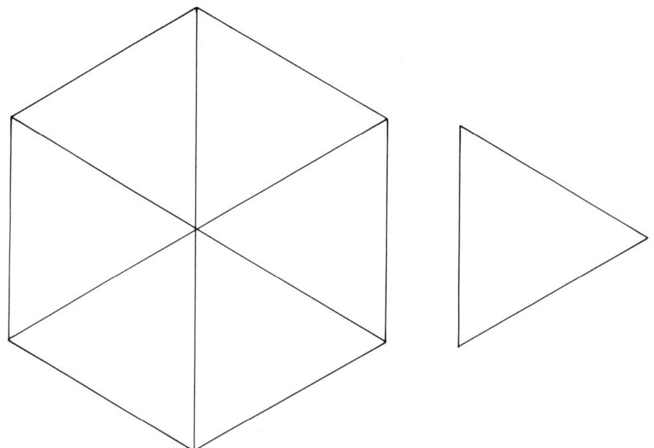

Fig. 37 (Model 6, p. 28)

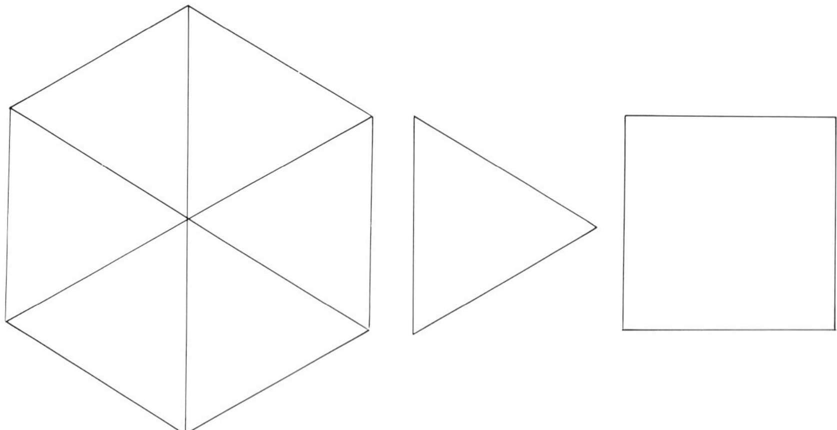

Fig. 38 (Model 7, p. 28)

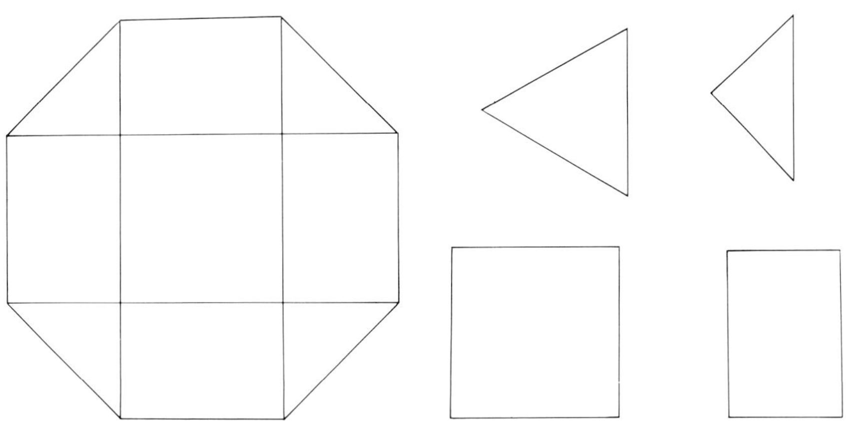

Fig. 39 (Model 4, p. 28)

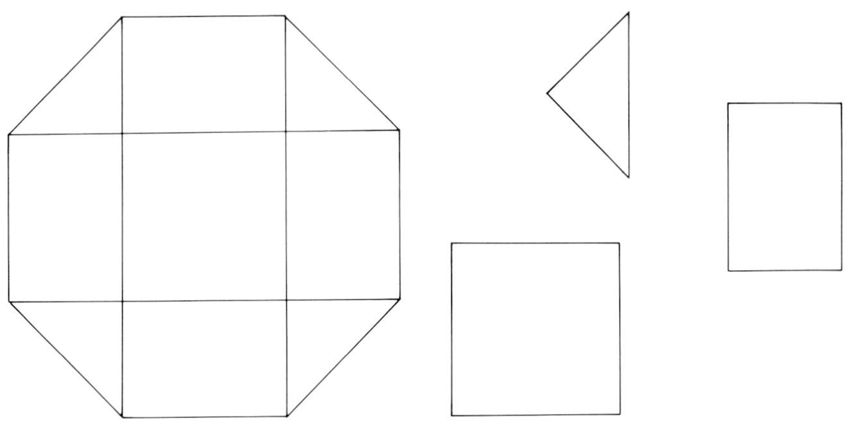

Fig. 40 (Model 5, p. 28)

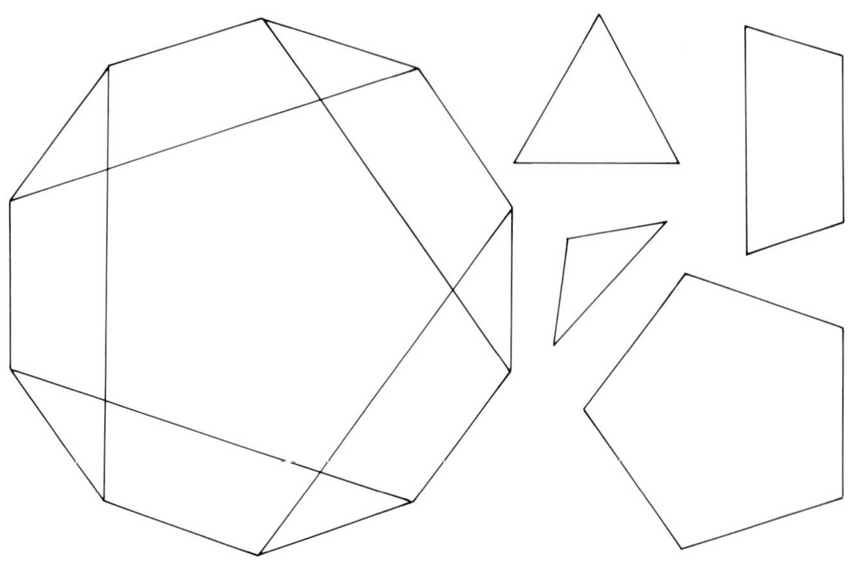

Fig. 41 (Model 8, p. 28)

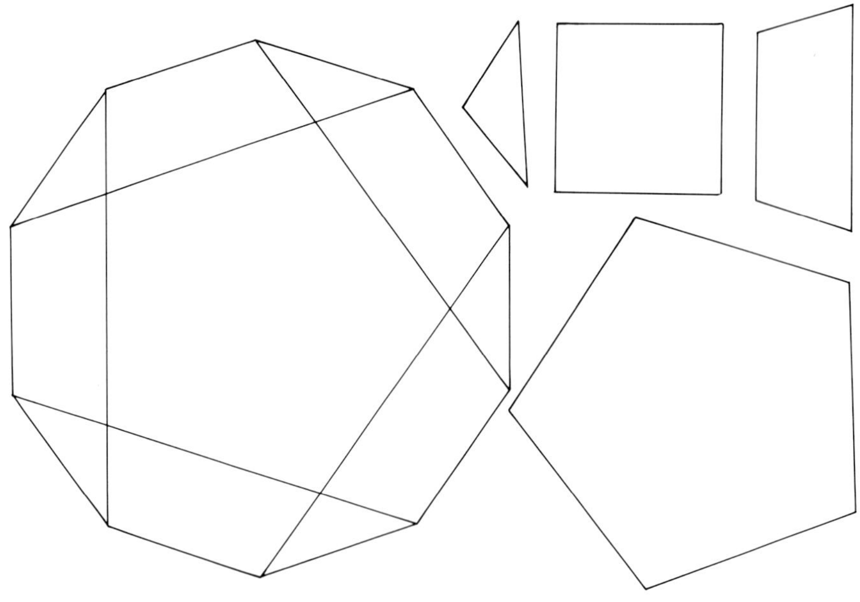

Fig. 42 (Model 9, p. 28)

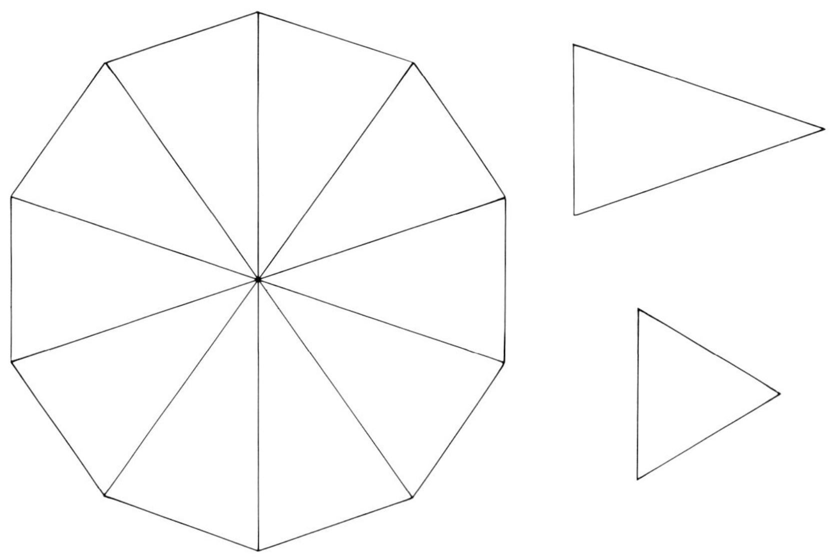

FIG. 43 (Model 3, p. 28)

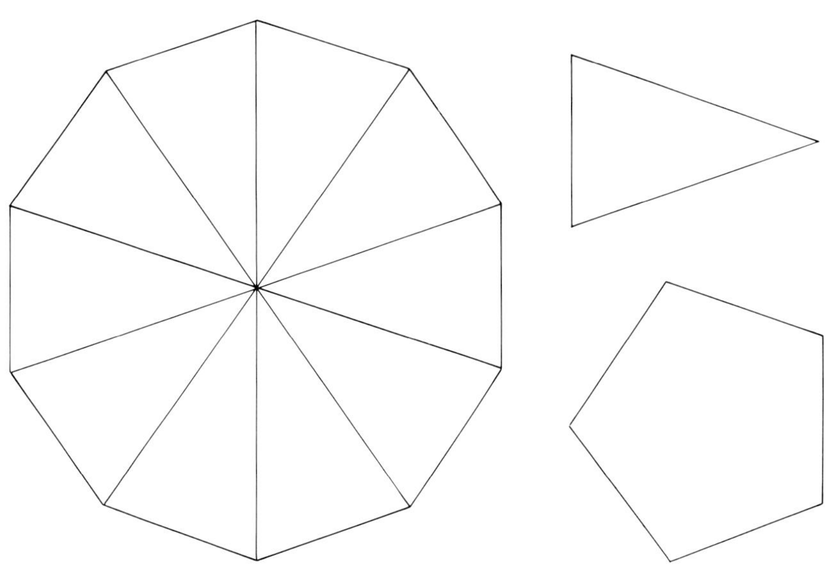

FIG. 44 (Model 2, p. 28)

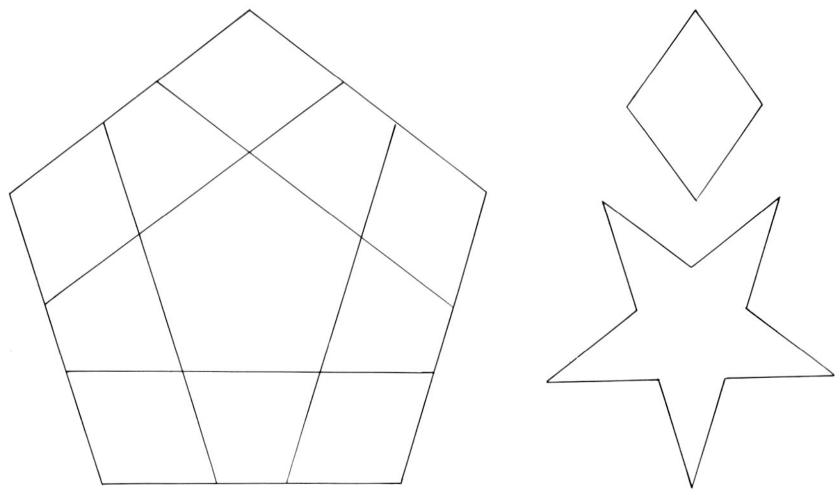

FIG. 45 (Model 10, p. 29)

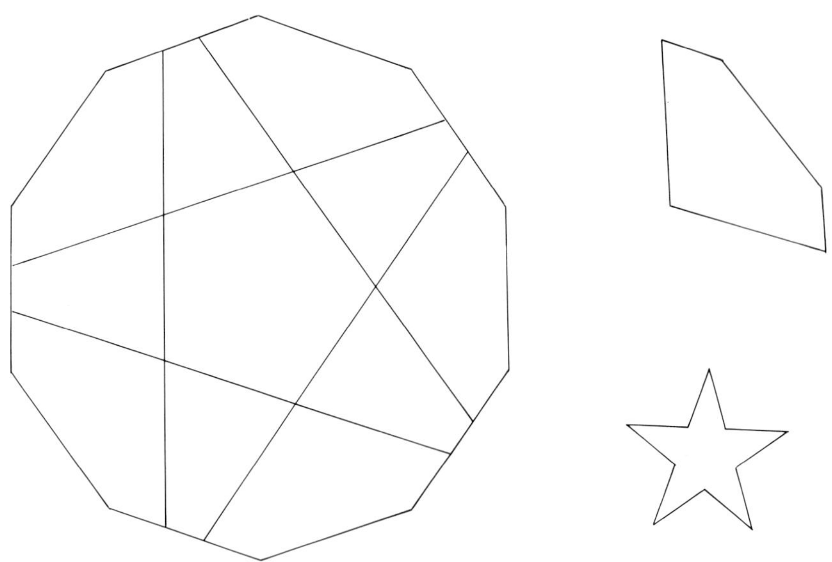

FIG. 46 (Model 11, p. 29)

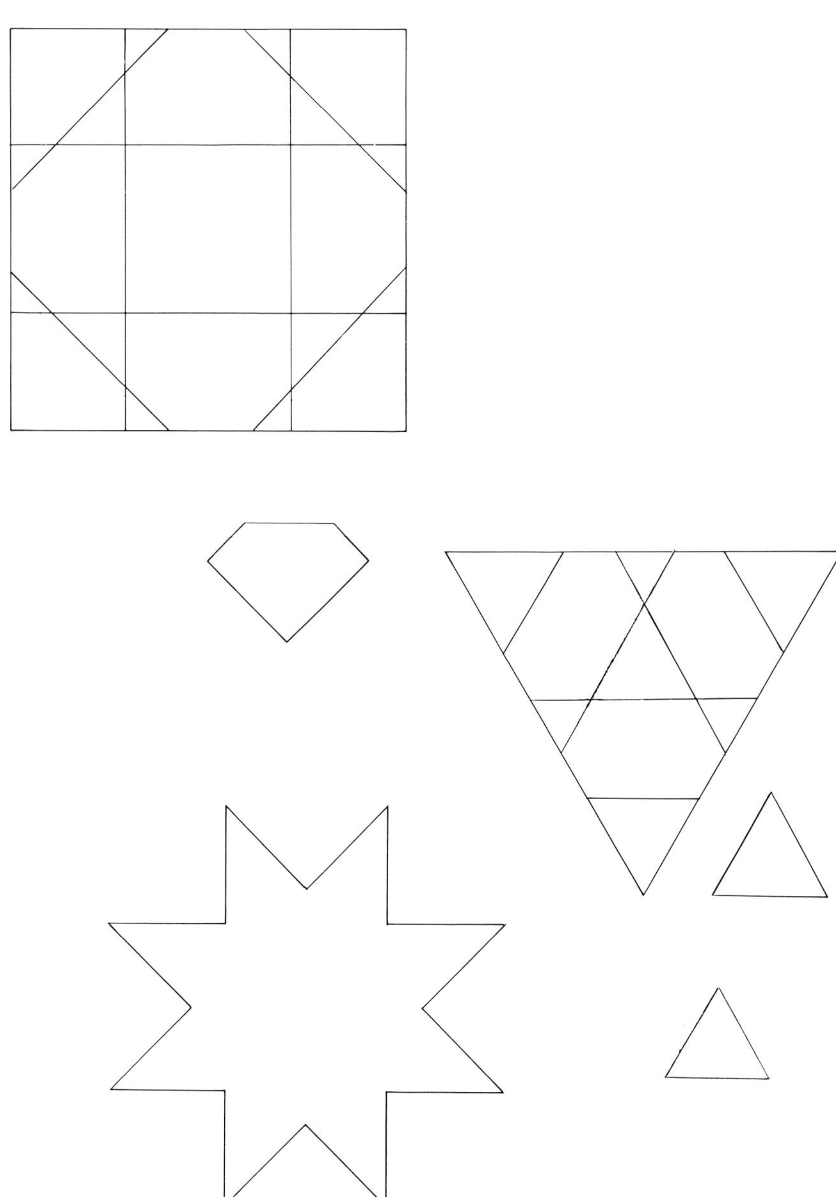

Fig. 47 (Model 12, p. 29)

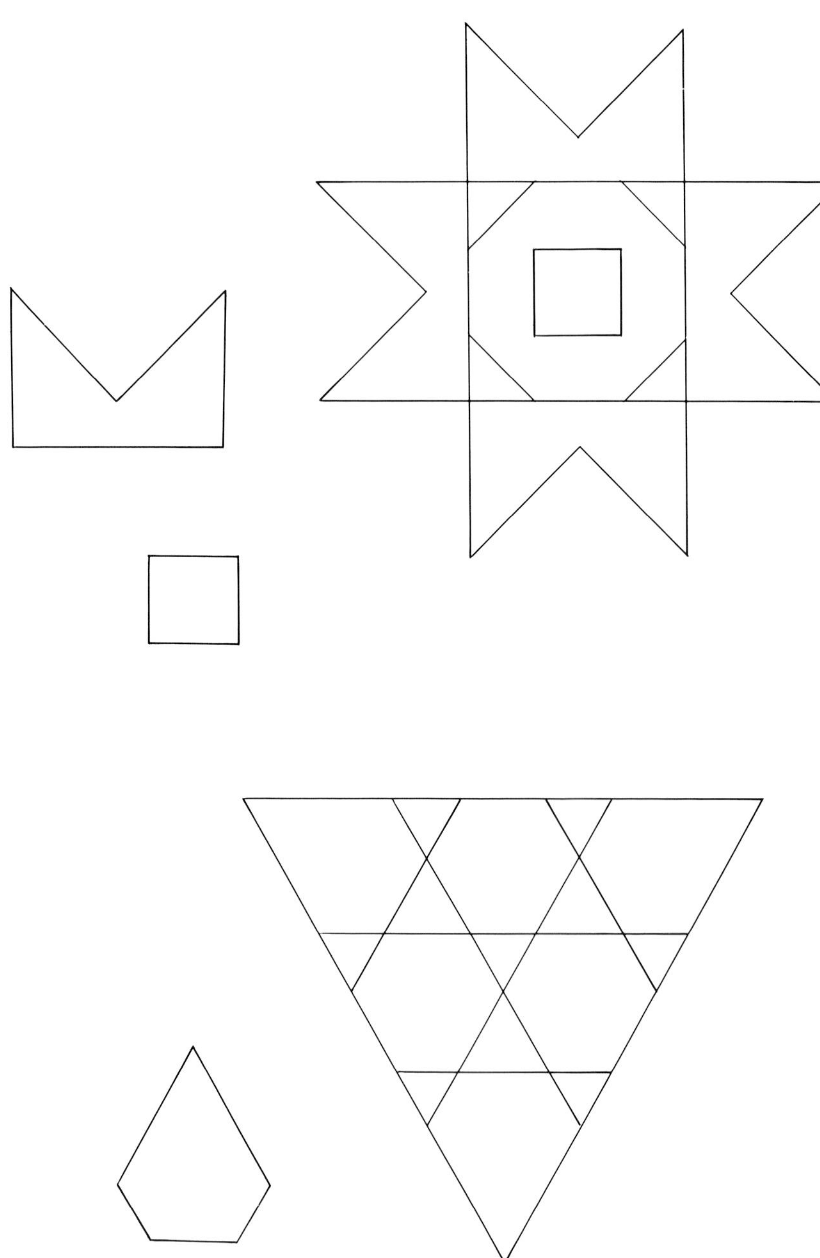

Fig. 48 (Model 13, p. 29)

FIG. 49 (Model 14, p. 29)

Fig. 50 (Model 15, p. 29)

Conclusion

Does the construction of polyhedron models have any practical significance apart from the hobby value it undoubtedly possesses? To answer this question, you may refer to Coxeter's book *Regular Polytopes*. To quote Coxeter: "The chief reason for studying regular polyhedra is still the same as in the time of the Pythagoreans, namely, that their symmetrical shapes appeal to one's artistic sense."[8] Again: "Anyone who believes that mathematics should be useful as well as beautiful should remember that polytopes have applications not only to the geometry of numbers but also to such practical subjects as the theory of communications and linear programming."[9]

In a geometry classroom polyhedron models may be used to illustrate the ideas of symmetry, reflection, rotation, and translation. Felix Klein, as far back as 1884, gave lectures on the regular solids and the theory of groups. His lectures on the icosahedron have thrown new light on the general quintic equation.[10] The various color arrangements suggested in this monograph might well be subjected to the mathematical analysis of group theory. At any rate, there is plenty of pure mathematics in the theory of polyhedra. More research may yet yield applications that up to now have never been thought possible.

Notes

1. Thomas L. Heath, *A History of Greek Mathematics* (New York: Oxford University Press, 1921), pp. 159-60.

2. H. S. M. Coxeter, M. S. Longuet-Higgins, and J. C. P. Miller, *Uniform Polyhedra* ("Philosophical Transactions of the Royal Society of London," Ser. A, Vol. CCXLVI, No. 196 [London: Cambridge University Press, 1954]), p. 402.

3. Heath, *op. cit.*, p. 162.

4. H. S. M. Coxeter, P. DuVal, H. T. Flather, and J. F. Petrie, *The Fifty-nine Icosahedra* ("Mathematical Series," No. 6 [Toronto: University of Toronto Press, 1938]), pp. 8-18.

5. Coxeter, Longuet-Higgins, and Miller, *op. cit.*, pp. 401-50.

6. Max Brueckner, *Vielecke und Vielfläche* (Leipzig: Teubner, 1900).

7. Coxeter, Longuet-Higgins, and Miller, *op. cit.*, p. 402.

8. H. S. M. Coxeter, *Regular Polytopes* (1st ed.; London: Methuen & Co., 1948), p. ix.

9. *Ibid.* (2nd ed.; New York: The Macmillan Co., 1963), p. viii.

10. Felix Klein, *Lectures on the Icosahedron* (New York: Dover Publications, 1956), esp. chap. i, "The Regular Solids and the Theory of Groups."

Bibliography

BALL, W. W. R., and COXETER, H. S. M. *Mathematical Recreations and Essays.* New York: The Macmillan Co., 1956. Chap. v, "Polyhedra."

BRUECKNER, MAX. *Vielecke und Vielfläche.* Leipzig: Teubner, 1900.

COXETER, H. S. M. *Regular Polytopes* (2nd ed.). New York: The Macmillan Co., 1963.

COXETER, H. S. M., DUVAL, P., FLATHER, H. T., and PETRIE, J. F. *The Fifty-nine Icosahedra.* ("Mathematical Series," No. 6.) Toronto: University of Toronto Press, 1938.

COXETER, H. S. M., LONGUET-HIGGINS, M. S., and MILLER, J. C. P. *Uniform Polyhedra.* ("Philosophical Transactions of the Royal Society of London," Ser. A, Vol. CCXLVI, No. 196.) London: Cambridge University Press, 1954. [This book can be ordered from The Royal Society, Burlington House, London W. 1, England.]

CUNDY, H. M., and ROLLETT, A. P. *Mathematical Models* (2nd ed.). New York: Oxford University Press, 1961.

LINES, L. *Solid Geometry.* New York: Dover Publications, 1965. Chap. xi, "Polyhedra"; chap. xii, "Semiregular and Star Polyhedra."

LYUSTERNIK, L. A. *Convex Figures and Polyhedra.* Translated by T. JEFFERSON SMITH. New York: Dover Publications, 1963.

STEINHAUS, H. *Mathematical Snapshots* (2nd ed). New York: Oxford University Press, 1960. Chap. vii, "Cubes, Spiders, Honeycombs, and Bricks"; chap. viii, "Platonic Solids, Crystals, Bees' Heads, and Soap."

For a more extensive bibliography, see

SCHAAF, WILLIAM L. *Recreational Mathematics: A Guide to the Literature.* Washington, D.C.: National Council of Teachers of Mathematics, 1958. Chap. iii, sec. 4, "Regular Polygons and Polyhedra," pp. 57-60.